T0239433

SpringerBriefs in Applied Sciences and Technology

SpringerBriefs present concise summaries of cutting-edge research and practical applications across a wide spectrum of fields. Featuring compact volumes of 50 to 125 pages, the series covers a range of content from professional to academic.

Typical publications can be:

- A timely report of state-of-the art methods
- An introduction to or a manual for the application of mathematical or computer techniques
- A bridge between new research results, as published in journal articles
- A snapshot of a hot or emerging topic
- An in-depth case study
- A presentation of core concepts that students must understand in order to make independent contributions

SpringerBriefs are characterized by fast, global electronic dissemination, standard publishing contracts, standardized manuscript preparation and formatting guidelines, and expedited production schedules.

On the one hand, **SpringerBriefs in Applied Sciences and Technology** are devoted to the publication of fundamentals and applications within the different classical engineering disciplines as well as in interdisciplinary fields that recently emerged between these areas. On the other hand, as the boundary separating fundamental research and applied technology is more and more dissolving, this series is particularly open to trans-disciplinary topics between fundamental science and engineering.

Indexed by EI-Compendex, SCOPUS and Springerlink.

More information about this series at http://www.springer.com/series/8884

Ajit Kumar Saxena · Amit Kumar

Fish Analysis for Drug and Chemicals Mediated Cellular Toxicity

Ajit Kumar Saxena
Department of Pathology
and Laboratory Medicine
All India Institute of Medical Sciences
Patna, Bihar, India

Amit Kumar
BioAxis DNA Research Centre Private Ltd.
Hyderabad, Telangana, India

ISSN 2191-530X   ISSN 2191-5318   (electronic)
SpringerBriefs in Applied Sciences and Technology
ISBN 978-981-15-4699-0   ISBN 978-981-15-4700-3   (eBook)
https://doi.org/10.1007/978-981-15-4700-3

This Springer imprint is published by the registered company Springer Nature Singapore Pte Ltd.
The registered company address is: 152 Beach Road, #21-01/04 Gateway East, Singapore 189721, Singapore

Contents

About the Authors

Prof. Ajit Kumar Saxena has received his Ph.D. from the Department of Anatomy in Cyotgenetics from the Institute of Medical Sciences, BHU, in the year of 1990. During his Ph.D., he studied how anticancer drugs cause interference with normal development of the fetus and with special reference to the cause of abnormality during differentiation of male gonads which lead to infertility including congenital anomalies. After receiving his Ph.D. degree, he joined NIH-funded Indo-US project at AIIMS New Delhi, where he identified the role of novel antigen IL6 as a signal traducing agent in human glioblastoma. He also worked in various renowned institutions of India and abroad including CCMB, CDRI, A.M.C., B.P.K.I.H.S. (Dharan, Nepal) and BHU, Varanasi. Currently, he is working as Professor and Head in the Department of Pathology/Lab Medicine specialized in clinical genetics in All India Institute of Medical Sciences Patna. He also possess administrative responsibilities as Head of the Department in prestigious institutions of India and abroad. In addition, he has also served his duties of Scientific Advisor to the government. His research interest includes study of gene polymorphism in congenital neural tube defects/tumor biology/infertility and cytokine gene expression in chronic wounds and hypertrophic scars. Currently, his research interest includes role of stem cells in translational research.

Of course, he published more than 100 articles in peer-reviewed journals of international reputes with high citations. Based on his research and fellowship training, he has received several awards and honors, such as Gold Medal Award, Confer 'Vivek Ratan Award' and Millennium Award, Health Care Excellence Award and Life Time Achievement Awards (three times). Besides his research, teaching and administration, he has also guided postgraduate (Ph.D., MD, MS, M.Tech.) students. He is serving as Editorial Member of various reputed journals and Reviewer for several government-funded projects and research papers. He also acquired and completed several government-funded research projects (ICMR, DBT, CSIR, DST). He is 'Fellow' and 'Member' of various scientific organizations, National Academy of Sciences, USA, and National Academy for Advancement of Science and Technology, USA, and Member of National Academy of Medical Sciences, India. He is also Fellow of Indian Association of Biomedical Scientist, Chennai, and Fellow of Indian Association of Applied Biotechnology, Mysore. Currently, he is engaged in the developing Advanced Clinical Genetics Laboratory for Diagnostic, Training and Research in India (Bihar).

Dr. Amit Kumar is a DNA Forensics Professional, Entrepreneur, Engineer, Bioinformatician and an IEEE Volunteer. In 2005 he founded first Private DNA Testing Company Bio Axis DNA Research Centre (P) Ltd in Hyderabad, India with an US Collaborator. He has vast experience of training 1000+ Crime investigation officers and helped 750+ Criminal and non-criminal cases to reach justice by offering analytical services in his laboratory. His group also works extensively on Genetic Predisposition risk studies of cancers and has been helping many cancer patients from 2012 to fight and win the battle against cancer. Amit was member of IEEE Strategy Development and Environmental Assessment committee (SDEA) of IEEEMGA. He is senior member of IEEE and has been a very active IEEE Volunteer at Section, Council, Region, Technical Societies of Computational Intelligence and Engineering in Medicine and Biology and at IEEE MGA levels in several capacities. He has driven number of IEEE Conferences, Conference leadership programs, Entrepreneurship development workshops, Innovation and Internship related events. Currently He is a Professor at SJB Institute of Technology Bangalore, IEEE MGA Nominations and Appointments committee member and Chairman of IEEE Hyderabad Section.

Chapter 1
Introduction of Cyclophosphamide

Recent epidemiological investigation in early pregnancy has revealed an increased embryonic loss in nurses occupationally exposed to antineoplastic drugs, during the first trimester of pregnancy [1]. Drug treatment often in pregnant female has been associated with a number of malformations in both animals [2, 3] and humans [4]. There has been also increasing number of reports of gonadal damage following cytotoxic drug therapy for the treatment of malignant and non-malignant conditions [5].

Drugs in the alkylating agent category used in cancer therapy, such as cyclophosphamide (CP), have also been anecdotally associated with gonadal damage in anomalies. Thus, the interest in the possible use of CP in therapeutic areas other than cancer chemotherapy has increased concern regarding the effects on reproductive processes [6]. In human pregnancy, it has been particularly difficult to obtain data on the action of drugs and environmental chemicals [7]. The use for an animal study in this direction therefore becomes imperative. Interestingly, what effect different doses of CP have on the male gonad of the offspring is exposed antenatally to CP during the critical period of gonadal development.

1.1 Structure of Cyclophosphamide

CP is a potent cytostatic agent related to the class of compounds known as alkylating agents. CP together with a number of related N-phosphorylated derivatives of nitrogen mustard is an attempt to obtain a compound with greater antitumor specificity than found in conventional alkylating agents of the nitrogen mustard class. Endoxan-Asta brand of cyclophosphamide is chemically N, N-bis (B-chloroethyl)-N-O-propylenephosphoric acid ester diamide monohydrate (Fig. 1.1) to release the active polyfunctional alkylating form.

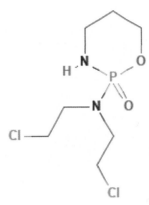

Fig. 1.1 Cyclophosphamide is a synthetic alkylating agent chemically related to the nitrogen mustards with antineoplastic and immunosuppressive activities. In the liver, cyclophosphamide is converted to the active metabolites aldophosphamide and phosphoramide mustard, which bind to DNA, thereby inhibiting DNA replication and initiating cell death, MW: 261.08 g/mol

1.2 Activation of Cyclophosphamide

The activity of CP in experimental and clinical cancer was demonstrated and now well established that the effectiveness of CP does not arise from enzymatic cleavage by tumor cells as originally proposed. This compound is activated by hepatic microsomal enzymes, with the active metabolites reaching their target site through the systemic circulation. The complex mode of enzymes with the active metabolites is reaching their target site through the systemic circulation. The complex mode of activation of CP and the identity of the active metabolites have been the subject of intense study by several groups of investigators. The current view of activation [8, 11], which may well undergo full further modification, is that the liver cytochrome p-450 microsomal mixed functional oxidase system converts cyclophosphamide to 4-hydroxycyclophospamide, within equilibrium with its acyclic tautomeric form, aldophosphamide.

Cyclophosphamide further oxidation of the latter compound possibly through the mediation of liver aldehyde oxidase and the other aldehyde-metabolizing enzymes results in the formation of known metabolites, carboxyphosphamide and 4-ketocyclophoasphamide which are relatively non-toxic. The cleavage of aldophosphamide by a ß-elimination reaction yields phosphoramide mustard and acrolin, both of which are highly cytotoxic and may represent the active form of the drug [9]. Although both of these compounds have now been isolated as products of cyclophosphamide activation [10, 11], it has not been established that they are generated in sufficient quantity to account for the total cytotoxic activity. Hydrolysis of P-N linkage of the parent compound or of its metabolites could yield the alkylating agent non-nitrogen mustard, also an end product of cyclophasphamide metabolites. The cytotoxic activity of this compound appears to be too low to contribute appreciably to the total pharmacologic activity of cyclophosphamide.

1.3 Risk Factors of Cyclophosphamide

CP is mutagenic [3], carcinogenic [12] and teratogenic [3, 18] in nature [13]. CP is a first substance known to induce chromosome rearrangements and gene mutation in germ cells of experimental animals [14]. The mutagenic activity has been reported for a variety of mammalian (including human) somatic cells and for germ cells of experimental mammals [14]. An increased number of chromosomal aberrations were observed in the peripheral blood lymphocytes of children treated with 3–5 mg/day of CP for 6–8 months for non-malignant conditions. Treatment with CP has been associated with the production of secondary cancer, significant incidence of bladder carcinoma, non-Hodgkin's lymphoma and squamous cell carcinoma of the skin which have been observed in patients treated for non-malignant diseases, rheumatoid arthritis and glomerulonephritis [15]. The incidence of a second cancer also increased in CP-treated cancer patients. These malignancies have been bladder cancer, acute leukemias and reticulum cell sarcomas in some unusual locations [16].

Teratogenic potential of CP in humans has been difficult to evaluate since there are no epidemiologic data to correctly assess the embryologic risk in man [16]. However, CP has been reported to cause gonadal damage (oligozoospermia, azoospermia and amenorrhoea) which may be irreversible in some instances [17]. Also pathologically in man, marked aplasia of sertoli's cells and loss of germinal epithelium, lining the seminiferous tubules are reported to be the striking features. In woman, ovarian atrophy, fibrosis and complete absence of follicular structures have further been observed to be the primary histologic features. Several reports have shown that high doses of CP on specific critical days of pregnancy can be teratogenic to experimental animals [18]. It has been suggested that CP is about three times more lethal to the fetus than the adult rats [19]. Large doses of CP have been reported to cause alopecia, nausea and vomiting [20]; fetal cardiomyopathy, hematologic effects, pulmonary toxicity [21]; hepatotoxicity [22]; mucosal ulceration, skin pigmentation and anaphylaxis. Transient cerebral dysfunction has also been associated with CP therapy.

CP has also been used an immunosuppressive agent following organ transplantation experiments [23]. It caused significant depression of antibody production with number of antigens [24]. CP was found to be more effective against rapidly proliferative cells. Since CP is more effective against proliferative (in cycle) than non-proliferative (out of cycle) cells, it was grouped 'cell cycle-specific' antineoplastic agents [25]. The cell cycle specificity of CP appears to be prime determinant in killing of immunologically competent cells. Embryo transplantation experiments showed that early CP treatment interfered with the subsequent development of both the embryo and the mother [26].

Objectives of study are based on existing knowledge in literature where CP has been used to see its effect on the adults, especially during spermatogenesis [14, 27–30]. Despite the fact that CP has been associated with gonadal damage in several reports, its action during differentiation and development of the testes has failed to receive even cursory attempt. It is undoubtedly to know the changes in testicular tissue collected from 1-day-old pups delivered to mother rats exposed to single dose, 2 or 10 or 20 mg/kg of CP on 12th or 15th or 18th day of gestational age to evaluate the genotoxicity in terms of chromosomal abnormalities and to compare the same in other organ such as liver and bone-marrow, protein profile alterations, and cellular changes based on histopathology. Besides this apart from the major objective, it was also proposed to evaluate the frequency of chromosomal aberrations in embryonic tissue and/or liver and testes at various intervals (14th, 16th, 18th and 22nd day of gestation), after single dose (20 mg/kg) of CP at 12th day of gestation. Further extended the study of the protein profile changes in normal developing embryos between 12th and 15th day of gestation, critical period when the embryos were insulted as a part of the major objective in this study, and to look into the reproductive performance (by breeding/mating experiment) followed by looking into the protein profile, in case infertility was induced, of the adult males and females from the following experiments where mother rats were exposed to single dose of CP of 2 or 10 or 20 mg/kg at 12th or 15th or 18th day of gestation, or a continuous dose (5 mg/kg) of CP on 12th to 15th day of gestation.

References

1. S.G. Selevan, M.L. Lindbohm, R.W. Hornung, K. Hemminki, A study of occupational exposure to antineoplastic drugs and fetal loss in nurses. N. Engl. J. Med. **313**(19), 1173–1178 (1985)
2. J.E. Gibson, B.A. Becker, The teratogenicity of cyclophosphamide in mice. Cancer Res. **28**(3), 475–480 (1968)
3. B.F. Hales, Modification of the mutagenicity and teratogenicity of cyclophosphamide in rats with inducers of the cytochromes P-450. Teratology **24**(1), 1–11 (1981)
4. L.H. Greenberg, K.R. Tanaka, Congenital anomalies probably induced by cyclophosphamide. JAMA **188**, 423–426 (1964)
5. S.M. Shalet, Effects of cancer chemotherapy on gonadal function of patients. Cancer Treat. Rev. **7**(3), 141–152 (1980)
6. J.A. Botta Jr., H.C. Hawkins, J.H. Weikel Jr., Effects of cyclophosphamide on fertility and general reproductive performance of rats. Toxicol. Appl. Pharmacol. **27**(3), 602–611 (1974)
7. R. Vogel, H. Spielmann, Potentiating effect of caffeine on embryotoxicity of cyclophosphamide treatment in vivo during the preimplantation period. Teratog. Carcinog. Mutagen. **7**(2), 169–174 (1987)
8. T.A. Connors, P.J. Cox, P.B. Farmer, A.B. Foster, M. Jarman, Some studies of the active intermediates formed in the microsomal metabolism of cyclophosphamide and isophosphamide. Biochem. Pharmacol. **23**(1), 115–129 (1974)
9. H.L. Gurtoo, S.K. Bansal, Z. Pavelic, R.F. Struck, Effects of the induction of hepatic microsomal metabolism on the toxicity of cyclophosphamide. Br. J. Cancer **51**(1), 67–75 (1985)
10. R.A. Alarcon, J. Meienhofer, Formation of the cytotoxic aldehyde acrolein during in vitro degradation of cyclophosphamide. Nat. New Biol. **233**(42), 250–252 (1971)

11. M. Colvin, C.A. Padgett, C. Fenselau, A biologically active metabolite of cyclophosphamide. Cancer Res. **33**(4), 915–918 (1973)

12. D. Schmähl, M. Habs, Carcinogenic action of low-dose cyclophosphamide given orally to Sprague-Dawley rats in a lifetime experiment. Int. J. Cancer **23**(5), 706–712 (1979)

13. J.M. Trasler, B.F. Hales, B. Robaire, Chronic low dose cyclophosphamide treatment of adult male rats: effect on fertility, pregnancy outcome and progeny. Biol. Reprod. **34**(2), 275–283 (1986)

14. G.R. Mohn, J. Ellenberger, Genetic effects of cyclophosphamide, ifosfamide and trofosfamide. Mutat. Res. **32**(3–4), 331–360 (1976)

15. L.J. Kinlen, R.N. Hoover, Lymphomas in renal transplant recipients: a search for clustering. Br. J. Cancer **40**(5), 798–801 (1979)

16. A.R. Ahmed, S.M. Hombal, Cyclophosphamide (Cytoxan). A review on relevant pharmacology and clinical uses. J. Am. Acad. Dermatol. **11**(6), 1115–1126 (1984)

17. N. Brock, The oxazaphosphorines. Cancer Treat. Rev. **10**(Suppl A), 3–15 (1983)

18. J.E. Gibson, B.A. Becker, Impairment of motor function of neonatal chicks by hemicholinium treatment during incubation. Toxicol. Appl. Pharmacol. **14**(2), 380–392 (1969)

19. S. Chaube, W. Kreis, K. Uchida, M.L. Murphy, The teratogenic effect of 1-beta-D-arabinofuranosylcytosine in the rat. Protection by deoxycytidine. Biochem. Pharmacol. **17**(7), 1213–1216 (1968)

20. J.H. Fetting, L.B. Grochow, M.F. Folstein, D.S. Ettinger, M. Colvin, The course of nausea and vomiting after high-dose cyclophosphamide. Cancer Treat. Rep. **66**(7), 1487–1493 (1982)

21. R.B. Weiss, F.M. Muggia, Cytotoxic drug-induced pulmonary disease: update 1980. Am. J. Med. **68**(2), 259–266 (1980)

22. A.M. Bacon, S.A. Rosenberg, Cyclophosphamide hepatotoxicity in a patient with systemic lupus erythematosus. Ann. Intern. Med. **97**(1), 62–63 (1982)

23. T.E. Starzl, J. Corman, C.G. Groth et al., Personal experience with orthotopic liver transplantation. Transplant. Proc. **4**(4), 759–771 (1972)

24. J.L. Turk, D. Parker, Effect of cyclophosphamide on immunological control mechanisms. Immunol. Rev. **65**, 99–113 (1982)

25. F. Valeriote, L. van Putten, Proliferation-dependent cytotoxicity of anticancer agents: a review. Cancer Res. **35**(10), 2619–2630 (1975)

26. R.S. Spielman, J.V. Neel, F.H. Li, Inbreeding estimation from population data: models, procedures and implications. Genetics **85**(2), 355–371 (1977)

27. I.P. Lee, R.L. Dixon, Effects of procarbazine on spermatogenesis determined by velocity sedimentation cell separation technique and serial mating. J. Pharmacol. Exp. Ther. **181**(2), 219–226 (1972)

28. K. Fukutani, H. Ishida, M. Shinohara et al., Suppression of spermatogenesis in patients with Behçet's disease treated with cyclophosphamide and colchicine. Fertil. Steril. **36**(1), 76–80 (1981)

29. F. Pacchierotti, D. Bellincampi, D. Civitareale, Cytogenetic observations, in mouse secondary spermatocytes, on numerical and structural chromosome aberrations induced by cyclophosphamide in various stages of spermatogenesis. Mutat. Res. **119**(2), 177–183 (1983)

30. R.L. Schilsky, B.J. Lewis, R.J. Sherins, R.C. Young, Gonadal dysfunction in patients receiving chemotherapy for cancer. Ann. Intern. Med. **93**(1), 109–114 (1980)

Chapter 2
Effect of Cyclophosphamide on Chromosomes

The chromosome damage has long been recognized as a serious hazard to man because such a damage is associated with severe clinical disorders [1–3]. Thus, tests designed to detect the chromosome-damaging potential (clastogenicity) of chemicals have become an established part of the safety evaluation programmes. The best validated in vivo test for clastogenic potential involves the analysis of either chromosomes or micronuclei in rodent bone-marrow cells [4]. The bone marrow is a useful source because it has proliferating cells with a short cell cycle and is recommended for estimating the mutagenic activity of chemicals [5, 6].

Cyclophosphamide (CP) is a first substance known to induce chromosome rearrangements and gene mutation in germ cells of experimental animals (see [7]). The clastogenic effect of CP has been assessed in human lymphocyte cultures [6, 8, 9] and bone-marrow cells. It is also reported to induce chromosomal aberrations and (SCEs) in human and experimental mammals in vivo [10–17]. Most of the chemical mutagens induce chromosomal aberrations both in bone-marrow cells and in spermatogonia. Besides equal or different sensitivities of various tissues [18], spermatogonia seem to be more sensitive to cell killing than bone-marrow cells. It was observed that single dose of CP (40 mg, 80 mg or 160 mg/kg) induced structural chromosomal aberrations including translocation and interfered with the normal development of bivalents [19]. Organ-specific genotoxicity effect of CP was observed by Ashby and Beije [20]. CP induced chromosomal aberrations and micronuclei in rat bone-marrow cells but failed to induce unscheduled DNA synthesis in liver [3, 20]. CP damage in bone-marrow cells of non-hepatectomized rats was compared with the CP damage on regenerating liver cells of hepatectomized rats. Results indicated that regenerating liver cells were more sensitive than bone-marrow cells [21].

Cytogenetic analysis of bone marrow in mice revealed that during the long-term exposure (7–28 days), the percentage of aberrant cells remained stable but considerable individual variation in the frequency of aberrant cells was observed [5]. The chromosomal aberration in bone-marrow cells was influenced by the length of the exposure to mutagen. The frequent aberrations were chromatid and chromosome breaks. Chromatid exchanges were detected only during the first 24 h after exposure termination, and chromosome exchange (dicentric and ring chromosome) was seen very rarely [22]. In mice, earlier studies [19, 23–25] showed a significant increase of chromosomal aberration after a single dose (20, 40 and 60 mg/kg) of CP. The maximum frequency of aberrant cells occurred 12 and 18 h after 50 mg/kg of CP application [22], while cytogenetic analysis of cells processed after 24 h yields only about 60–70% of maximum frequency changes [26–28].

2.1 Experimental Procedure

Rats are putative animal being used for experimental studies. Female Charles Foster rats of approximately 100–200 days of age weighing about 200–250 g were housed in wire cages with the males of the same strain (C.F. inbred strain) in the evening, and pregnancy was confirmed by the presence of sperms in the vaginal smear examined on the following morning. Sperm-positive day was designated as day 'zero' of pregnancy. Pregnant females were weighed and houses individually in a noise-free air-conditioned (23 ± 1 °C) animal house maintained on a light dark cycle of 12:12 h. Pregnant animals were fed on the diet pellet (Hindustan Lever, Bombay, India) and tap water.

Chemical The chemicals used were cyclophosphamide (Endoxan-Asta) (Khandelwal Lab., Bombay, India); colchicine (Bios, India); sodium chloride; potassium chloride; trisodium citrate; acetic acid; methanol; glycerol; Giemsa powder (B.D.H., India); trypsin (Sigma Chemicals, Co., USA). The stock CP solution was prepared under the sterile conditions by dissolving 100 mg of CP in 10 ml of sterile distilled water, and the dosage of the drug was worked out on the basis of mg/kg body weight. Nine parallel sets of experiments were carried out, where one of the three different doses, i.e., 2 or 10 or 20 mg/kg body weight of CP was administered intraperitoneally (i.p.), with the help of a sterile tuberculin syringe to the pregnant rats at either of the three different, i.e., 12th or 15th or 18th day of gestation. These experiments were designed according to the dose/gestation as: 2/12, 10/12, 20/12; 2/15, 10/15, 20/15; and 2/18, 10/18, 20/18. The pregnancy in all these experimental groups was allowed to continue till delivery, and the tissues such as testes, liver and bone marrow collected from 1-day-old male pups for cytogenetic studies. In another set of experiments, single dose of 20 mg/kg body weight of cytophosphamide was given at 12th day of gestation after dissecting the uteri horn for cytogenetic studies. Similar procedure was adopted for collection of fetuses on 22nd day in the experiments with delayed delivery. The control animals were divided into two major groups: one which received the normal saline at three different gestations (12, 15, 18) parallel to the experimental groups and the other with no injection at all. Because of the absence of a significant difference between the two types of controls, the data was pooled and the average values are used for comparison in the study.

Metaphase arrest The colchicines, a mitotic inhibitor (8 mg/kg body weight), were injected to 1-day-old male pups, and pups sacrificed by cervical dislocation after two hours of exposure.

Chromosome preparation The fresh tissues from 1-day-old male pups and embryonic/fetuses were collected in a small glass petri dish or watch glass and minced with a pair of scissors where required. This tissue suspension in saline was transferred to the graduated conical centrifuge of the clumped cells and centrifuged at 800 rpm for 10 min and decanted.

Hypotonic treatment The pellet was suspended in prewarmed freshly prepared 0.56% potassium chloride solution or 0.9% sodium citrate solution in case of testicular suspension and kept at 37 °C for 10–20 min. The cells were dispersed by gentle pipetting and centrifuged at 800 rpm for 10 min.

Fixation The pellet thus obtained was fixed in 3:1 methanol: acetic acid. The fixative was added slowly and changed thrice before the slide was prepared.

Slide preparation The layer of dispersed cell suspension was drawn out, and three to four drops were added onto a precooled clean slide. The slide was then held close to the flame for few moments, for the fixative to burn completely, leaving the slide dry.

Chromosome staining and Giemsa banding The slide was stained with 5% Giemsa stain or used for G-banding. The G-banding was done by the method of Worton and Duff (1979), which is a modified technique of Seabright. The banding was done at room temperature, and slides are rinsed in 0.15 M NaCl and exposed to 0.15 M NaCl containing trypsin solution for 30–120 s. The slides were again rinsed with 0.15 M NaCl solution and then with 5% Giemsa for 5–10 min, followed by a wash in distilled water and allowed to dry. The slides were observed for cytogenetic features and photographed. The normal karyotype with 42 autosomes and X and Y sex chromosomes of male Charles Foster rats as shown in Fig. 2.1.

The chromosomal features such as breaks, gaps, dicentrics, acentric fragments and ring chromosomes were scored in unbanded metaphase plates as many times breaks in chromosomes were not discernible clearly in banded metaphases. The chromosome features observed in the study were divided into two groups: one including breaks, gaps, acentric fragments, dicentrics and ring chromosomes (Fig. 2.2) and the other including centromere spreading, chromatin bodies (Fig. 2.3) and aneuploidy. The breaks and gaps in the former group were distinguished, and dicentric chromosomes could not be confirmed by C-banding because most of the scoring of the features in this group was done in unbanded metaphase plates. This was due to the reason of missing breaks and gaps in swollen banded plates at times. The identification of such features of centromeric spreading includes in the second group. The name 'chromatin bodies' in this study are used as per convenience for fragmented chromosomes, where fragmented pieces round up in different sizes to form chromatin bodies. The probable mechanism of formation of such bodies and the end result is shown in Fig. 2.3A, B. This feature seems apparently to be different from pulverization of chromosomes. Further, the scoring of hyperdiploids (without exact multiples of basic diploid number) was done as aneuploidy. Hypodiploid metaphase plates were only

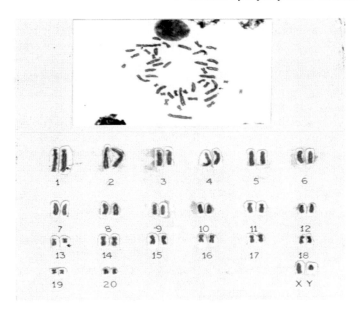

Fig. 2.1 Normal karyotype of male Charles Foster rats

Fig. 2.2 A–E Partial metaphase showing breaks, gaps, acentric fragments, dicentrics and ring chromosomes

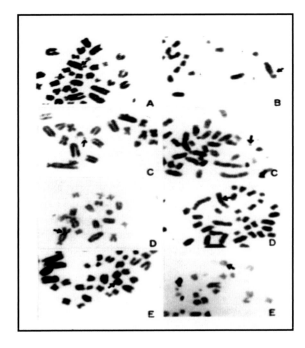

Fig. 2.3 A–D Partial metaphase showing centromere spreading (**C** and **D**) formation of chromatin bodies after extreme fragmentation of chromosome (**A** and **B**)

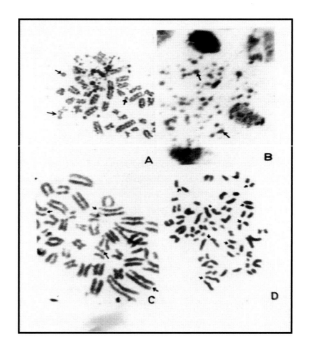

considered if consistently observed with the same chromosome number. The dose and days relationship between nine parallel sets of experiments (already described) was observed using arcsine transformation for proportions and a two-way analysis of variance (*F*-values) in all the three tissues, i.e., testes, liver and bone marrow. Further, chi-square test was applied to compare the differences between treated and control experiments with respect to different chromosomal abnormalities. The correlation coefficient '*r*' was measured to examine the relationship between the three different doses used and the aneuploidy in different chromosome groups. The correlation of frequency distribution of aneuploidy of each chromosome between two different groups was also measured and showing significant relationship.

A variety of abnormal chromosomal features were observed in testicular tissue, liver and bone marrow of 1-day-old pups, obtained from pregnant rats, where cyclophosphamide (CP) was administered in a signal dose of 2 or 10 or 20 mg/kg body weight either at 12th or 15th or 18th day of gestation. The abnormal features such as breaks, gaps, acentric fragments, dicentrics, ring chromatin, centromeric spreading, chromatin bodies and aneuploids were observed in three different organs, i.e., in tests, liver and bone marrow in the nine experiments designated according to dose/day (gestation) as 2/12, 10/12, 20/12, 2/15, 10/15, 20/15 and 2/18, 10/18, 20/18 as well as in their respective controls.

The apartment observations of the frequency of total abnormal chromosomal features both in testicular tissue and liver from 1-day-old pups showed a dose-dependent increase irrespective of the gestational age at which they were exposed antenatally to cyclophosphamide. A maximum genotoxic effect in terms of chromosomal abnormalities was observed not only in testes (31.03%) and liver (42.85%), but also in bone marrow (50%) of 1-day-old pups exposed antenatally to 20 mg/kg dose of cyclophosphamide at 12th day of gestation (Expt. 20/12). A statistical comparison of data of total chromosomal abnormalities in testis of 1-day-old pups between the antenatally treated and control showed significant difference ($p < 0.5$) in the experiments where 10 mg/kg dose of cyclophosphamide was used at 12th (Expt. 10/12) and 18th (Expt. 10/18) day of gestation, whereas 20 mg/kg dose of cyclophosphamide to the pregnant rats at all the three gestations (12th, 15th and 18th day, i.e., Expt. 20/12, 20/15, 20/18) showed a high significant difference ($p < 0.001$) in comparison to control experiments. A similar comparison in other two tissues, i.e., liver and bone marrow from antenatally exposed 1-day-old male pups (Expt. 20/12) where 20 mg/kg dose of CP was administered to pregnant rats at day 12 gestation, also showed a significant difference ($p < 0.001$) with respect to controls. However, F-values calculated to compare the days and dose response with respect to studied chromosomal abnormalities in all the tissues showed lack of significant difference (Table 2.1).

The data is split into two groups, one including chromosome breaks, gaps, dicentric, ring chromosome, acentric fragment (Tables 2.2, 2.3 and 2.4) and the other with centromere spreading. Chromatin bodies and aneuploidy (Tables 2.5, 2.6 and 2.7) were carried out in all the three tissues. The statistical comparison was made to find out which specific aberration(s) contributed to the significant difference observed between experimental groups and its respective controls.

An apparent dose-related increase of chromosomal abnormalities like breaks, gaps acentric fragments, dicentrics and ring chromosomes was observed in testicular tissues of antenatally exposed 1-day-old pups in 2/18 (4.40%).

10/18 (13.46%) and 20/18 (15.38%) experimental groups. However, a decrease in the percentage frequency of these chromosome aberrations and no relationship

Table 2.1 Observed f-values between the experimental groups with respect to chromosomal abnormalities

Abnormal chromosomal features	Liver		Bone marrow		Testes	
	Days	Doses	Days	Doses	Days	Doses
Acentric fragment	0.20	1.79	1.28	0.71	0.31	1.68
Aneuploid	1.00	1.00	1.00	1.00	0.76	1.23
Breaks	1.92	0.07	1.00	1.00	1.49	0.51
Centromeric spreading	1.91	0.08	1.48	0.51	0.42	1.57
Chromatin bodies	0.49	1.56	1.14	0.85	0.001	1.99
Dicentric	0.15	1.84	0.43	1.56	0.35	1.63
Gap	1.02	0.87	1.02	0.98	1.34	0.61
Ring chromosomes	0.00	0.00	1.91	0.086	0.68	0.31
Total abnormalities	0.58	1.41	0.77	1.22	0.47	1.52

f-value (table) 5–1%: 2 and 4 d. f. 6.94 18.00 (lack of significant)

Table 2.2 A group of chromosomal features observed in testicular tissue collected from 1-day-old pups exposed in utero to CP at different gestations

Gestation period (days)	Doses of CP (mg/kg)	Total metaphase plates obs.	Total % (%) abnormal plates obs.	Number and (%) of abnormal chromosomal features				
				Break	Gap	Dicentrics	Acentric	Ring chromosomes
Control	–	37						
12	2	50	6 (12.00)	1 (2.00)	2 (4.00)	3 (6.00)		
	10	71	5 (7.04)	5 (7.04)				
	20	58	4 (6.89)	1 (0.58)	1 (0.58)			
15	2	41	3 (7.31)		1 (2.40)	2 (4.81)	2 (3.44)	
	10	83	3 (3.61)	1 (1.20)	2 (2.40)			
	20	90	9 (10.09)	2 (2.22)	1 (1.11)	3 (3.33)	2 (2.22)	1 (1.11)
18	2	45	2 (4.44)		1 (2.22)	1 (2.22)		
	10	52	7 (13.46)		5 (9.61)	1 (1.92)	1 (1.92)	
	20	91	14 (15.38)	1 (1.09)	2 (2.19)	3 (3.29)	6 (6.59)	2 (2.19)

at all with increasing dose was observed in 2/12, 10/12, 20/12 and 2/15, 10/15, 20/15 experimental groups, respectively, whereas in liver in 2/15 (1.15%); 10/15 (4.30%), 20/15 (9.09%) and in bone marrow of 2/15 (2.08%), 10/15 (6.25%), 20/15 (8.62%) experimental groups, an apparent dose-related response was also observed. The statistical analysis carried out with respect to control in all the three tissues showed lack of significant difference.

The frequency of chromosomal abnormalities like centromeric spreading. Chromatin bodies and aneuploidy increased with increasing dose in all the nine sets of experiments in testicular tissue. A maximum frequency (24.13%) was observed in 20 mg/kg dose at 12th day of gestation (i.e., Expt. 20/12). The chi-square test carried out between 2/12; 2/15; 2/18 and the control was insignificant, whereas the statistical comparison between 10/12 and 10/18 experimental groups and the control showed a significant difference ($p < 0.05$). This difference was contributed by the two chromosomal features, i.e., centromeric spreading and aneuploidy. However, the highest dose of cyclophosphamide (i.e., 20 mg/kg) showed a significant difference ($p < 0.01$) when experimental groups 20/12, 20/15 and 20/18 were compared with control. It was observed that chromatin bodies contributed to this significant difference with respect to control. The percentage frequency of abnormal chromosomal features such

Table 2.3 A group of chromosomal features observed in liver collected from 1-day-old pups exposed in utero to CP at different gestations

Gestation period (days)	Doses of CP (mg/kg)	Total metaphase plates obs.	Total % (%) abnormal plates obs.	Number and (%) of abnormal chromosomal features				
				Break	Gap	Dicentrics	Acentric	Ring chromosomes
Control	–	32	1 (3.12)		1 (3.12)			
12	2	189	5 (2.64)	2 (1.05)	2 (1.05)	1 (0.52)		
	10	87	2 (2.29)		1 (1.14)		1 (1.14)	
	20	77	2 (2.59)			1 (1.29)	1 (1.29)	
15	2	396	6 (1.51)	1 (0.25)	2 (0.50)	2 (0.50)	1 (0.25)	
	10	186	8 (4.30)	6 (3.22)	2 (0.50)			
	20	77	7 (9.09)	1 (1.29)	1 (1.29)	3 (3.89)	2 (2.59)	
18	2	33	3 (9.09)				3 (9.09)	
	10	138	5 (3.62)		1 (0.72)	2 (1.44)	2 (1.44)	
	20	32	3 (9.37)			1 (3.12)	2 (6.25)	

as break, gap, acentric fragments dicentrics and ring chromosomes in different tissues (A) testes, (B) liver and (C) bone marrow in 1-day-old pups exposed in utero to CP (2 or 10 or 20 mg/kg) at 12th or 15th or 18th day of gestation.

The percentage frequency of abnormal chromosomal features like centromeric spreading, chromatin bodies and aneuploidy in different tissues—(A) testicular tissue, liver and (C) bone-marrow in 1-day-old pups exposed in utero with CP (2 or 10 or 20 mg/kg) at 12th or 15th or 18th day of gestation.

The liver showed dose-dependent increase of chromosomal abnormalities such as centromeric spreading, chromatin bodies and aneuploidy in 2/12 (0.52%), 10/12 (40.25%) experimental groups, whereas a similar observation was made in bone marrow of 2/18 (2.27%); 10/18 (3.22%) and 20/18 (14.81%) experimental groups. However, the genotoxicity was observed in 20/12 experimental group in both liver and bone marrow. The statistical comparison revealed that chromatin bodies contributed to the significant difference ($p < 0.001$) observed in both liver and bone-marrow cells in comparison to their respective controls.

The percentage frequency of total aneuploid metaphase plates and the frequency of aneuploid chromosomes within a given group in testicular tissue in all the nine sets of experiments are depicted in Table 2.8. The F-values calculated to compare the days and dose response between the experimental groups with respect to total

Table 2.4 A group of chromosomal features observed in bone marrow of 1-day-old pups exposed in utero to CP at different gestations

Gestation period (days)	Doses of CP (mg/kg)	Total metaphase plates obs.	Total % (%) abnormal plates obs.	Number and (%) of abnormal chromosomal features				
				Break	Gap	Dicentrics	Acentric	Ring chromosomes
Control		44	0					
12	2	30	2 (6.66)			1 (3.33)		1 (3.33)
	10	64	2 (3.12)				1 (1.56)	1 (1.56)
	20	18	1 (5.55)				1 (5.55)	
15	2	48	1 (2.08)		1 (2.08)			
	10	16	1 (6.25)		1 (6.25)			
	20	58	5 (8.62)		1 (1.72)	3 (5.17)		1 (1.72)
18	2	44	3 (6.81)			1 (2.27)	1 (4.54)	
	10	124	10 (8.06)	5 (4.03)	2 (1.61)	2 (1.61)	1 (0.80)	
	20	54	0					

aneuploidy showed no aneuploidy increased with increasing doses of CP in experimental groups of 2/12 (4%); 10/12 (12.67%); 10/15 (2.40%), 20/15 (7.77%), and frequency (12.67%) of aneuploidy was observed only in 10/12 experimental group.

Table 2.8 shows the percent frequency of aneuploidy of individual chromosomes within a given chromosome group in the testicular tissue of 1-day-old pups in all the sets of experiments as also shown in karyotype (Fig. 2.4).

Apparently, the percent frequency was observed to increase in 10/12 experimental groups in both C- and D-group chromosomes in comparison to other groups. A similar observation was again made in C- and D-group chromosomes in the 20/15 experimental group. The correlation coefficient 'r' between the three different doses used and the aneuploidy observed in C- and D-group chromosomes were found to be significant ($p < 0.05$). However, aneuploidy of individual chromosomes in C-group, in experimental groups 2/18, 10/18 and 20/18, showed a dose-related increase. The correlation coefficient 'r' between the two was found to be significant ($p < 0.05$). A significant difference ($p < 0.01$) was also observed when 'r' was calculated between C and D groups of chromosomes for aneuploidy in 10/12; 10/15 and 10/18 experimental groups. Further, the correlation 'r' between the frequency distributed of aneuploidy of each chromosome within C- and D-groups (i.e., how many times each chromosome—11–13 in C-group and 14–20 in D-group appears in single aneuploid metaphase plate) showed a significant difference ($p < 0.05$) only within D-group chromosomes in 10/12, 10/15 and 10/18 experimental groups.

Table 2.5 A group of chromosomal features observed in testicular tissue of 1-day-old pups exposed in utero to CP at different gestations

Gestation period (days)	Doses of CP (mg/kg)	Total metaphase plates obs.	Total % (%) abnormal plates obs.	Number and (%) of abnormal chromosomal features		
				Centromeric spreading	Chromatin bodies	Aneuploidy
Control		37	2 (5.40)			2 (5.40)
12	2	50	3 (6.00)	1 (2.00)		2 (4.00)
	10	71	12 (16.90)	3 (4.22)		9 (12.67)
	*20	58	14 (24.13)	1 (0.58)	10 (17.24)	3 (5.17)
15	2	41	1 (2.43)	1 (2.43)		
	10	83	9 (10.84)	1 (1.20)	6 (7.22)	2 (2.40)
	20	90	11 (12.22)	1 (1.11)	3 (3.33)	7 (7.77)
18	2	45	2 (4.44)		1 (2.22)	1 (2.22)
	10	52	6 (11.53)	3 (5.76)		3 (5.76)
	20	91	13 (14.28)	1 (1.09)	8 (8.79)	4 (4.39)

In another set of experiment where single dose (20 mg/kg) collected at various intervals (i.e., 14th, 16th, 18th and 22nd day of gestation), abnormal chromosomal features like breaks, gaps dicentrics, acentric fragments, chromatin bodies, aneuploidy and centromeric spreading were observed. The percent frequency of total abnormal chromosomal features in embryonic tissues collected from 14th (20.28%), 16th (29.28%) and 22nd (72.72%) day fetuses showed an increase in chromosomal abnormalities with increasing gestational age. The statistical comparison was carried out with respect to control, using X^2-test which showed significant difference for 14th day ($p < 0.05$); 16th day ($p < 0.01$) and a very high significant difference for 22nd day ($p < 0.001$) in comparison to respective controls. A similar observation of increase in total chromosomal abnormality with increasing gestation was observed in case of liver collected from 16th (23.13%), 18th (44.15%) and 22nd (88.50%) day fetuses. The statistical comparison with respect to control showed significant difference from 16th day ($p < 0.05$) and for 22nd day ($p < 0.001$) fetuses. The chromosomal frequency in testicular tissue (28.35%) collected from 18th day fetus also showed significant differences ($p < 0.05$) with respect to control.

Table 2.6 A group of chromosomal features observed in liver of 1-day-old pups exposed in utero to CP at different gestations

Gestation period (days)	Doses of CP (mg/kg)	Total metaphase plates obs.	Total % (%) abnormal plates obs.	Number and (%) of abnormal chromosomal features		
				Centromeric spreading	Chromatin bodies	Aneuploidy
Control		32	0			
12	2	189	1 (0.52)	1 (0.52)		
	10	87	2 (2.29)	2 (2.29)	29 (37.66)	1 (1.29)
	20	77	31 (40.25)	1 (1.29)	1 (0.25)	
15	2	396	4 (1.01)	3 (0.75)		
	10	186	0			
	20	77	0			
18	2	33	0			
	10	138	19 (13.76)	3 (2.17)		16 (11.59)
	20	32	4 (12.50)		4 (12.50)	

Further analysis of the data after splitting into two groups, one including chromosome breaks, gaps, acentric fragments, dicentrics, ring chromosomes and the other with centromeric spreading chromatin bodies and aneuploidy, was carried out in all the tissues. The percentage frequency of total abnormal chromosomal features in different tissue collected at various intervals (14, 16, 18 and 22 days of gestation) after single dose of (20 mg/kg) CP at day of gestation as well as different groups—one showing the percentage frequency of abnormal chromosomal features like break, gap, acentric fragment, dicentric and ring chromosome and centromeric spreading, chromatin bodies, and aneuploidy in different tissues.

The chromosome fragments, dicentrics and rings which apparently decreased with respect to gestational age in liver are collected from 16th (13.63%) to 18th (22nd) (0.55%) day fetuses and show lack of significant difference with respect to control in all the tissue. The frequency of chromosomal abnormalities like centromeric spreading, chromatin bodies and aneuploidy increases in embryonic tissue with increase of embryonic age, i.e., 14th (8.69%), 16th (19.28%), 18th and 22nd (72.72%) day fetuses. Similar observation was observed in liver collected from 16th (12.50%), 18th (31.16%) and 22nd (85.47%) day fetuses. The testicular tissue of 18th (25.37%) day fetus showed increased frequency of chromosomal abnormalities with respect to control. The statistical analysis was carried in all the tissue with respect to control

Table 2.7 A group of chromosomal features observed in bone marrow of 1-day-old pups exposed in utero to CP at different gestations

Gestation period (days)	Doses of CP (mg/kg)	Total metaphase plates obs.	Total % (%) abnormal plates obs.	Number and (%) of abnormal chromosomal features		
				Centromeric spreading	Chromatin bodies	Aneuploidy
Control		44	1 (2.27)		1 (2.27)	
12	2	24	3 (12.50)	1 (3.33)	2 (6.66)	
	10	64	3 (4.64)	2 (3.12)	1 (1.56)	
	*20	18	8 (44.46)		8 (44.46)	
15	2	48	2 (4.16)	2 (4.16)		
	10	16	2 (12.56)	1 (6.25)	1 (6.25)	
	20	18	3 (5.17)	3 (5.17)		
18	2	44	1 (2.27)		1 (2.27)	
	10	124	4 (3.22)	2 (1.61)	1 (0.08)	1 (0.80)
	20	54	8 (14.81)		8 (14.81)	

which showed significant difference. It was observed in experiments that chromatin bodies contributed to the significant difference with respect to control.

The two types of fetuses were collected: one phenotypically normal looking and other phenotypically abnormal looking with clinical features such as hydrocephalus, micrognathia, growth retardation (loss of body weight), limbs reduced, product tongue, kink tail and swell umbilical cord observed. The percent frequency of total chromosomal abnormalities in liver from normal looking fetus (38.73%) was lower than liver of abnormal fetus (58.13%) collected from 18th day of gestation. The statistical comparison was carried out with respect to control. Using X^2-test showed significant difference ($p < 0.001$), whereas similar observation was observed in case of liver collected from 22nd day of gestation. The percent frequency of total chromosomal abnormalities in case of normal looking fetus (42.85%) was lower than what was observed in the liver collected from abnormal looking fetus (88.50%). The statistical comparison was carried out with respect to control and total abnormalities in case of liver from normal looking and abnormal looking fetus showed significant difference ($p < 0.001$).

Table 2.8 Total number of aneuploid metaphase plates and their (%) frequency of aneuploid chromosomes within a given group

Gestation period (days)	Doses of CP (mg/kg)	Total metaphase plates obs.	Total and (%) abnormal metaphase plates obs.	Frequency of aneuploidy of individual chromosomes			
				A (1–3)	B (4–10 X, Y)	C* (11–13)	D* (14–20)
Control		37	2 (5.40)				
12	2	50	2 (4.00)	1	2	4	6
	10	71	9 (12.67)	4	4	13	20
	20	58	3 (5.17)	1		3	5
15	2	41	0				
	10	83	2 (2.40)		1	2	4
	20	90	7 (7.77)		2	11	13
18	2	45	1 (2.22)	1	1	2	4
	10	52	3 (5.76)	1	2	4	6
	20	91	4 (4.39)	1		11	5

*Statistical analysis showing significant differences ($p < 0.05$)

Fig. 2.4 Karyotype showing aneuploidy in testicular tissue after antenatal exposure with cyclophosphamide

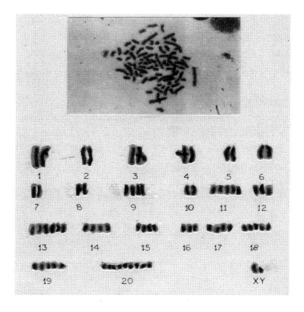

Further analysis of the data after splitting into two groups, one including chromosome breaks, gaps, acentric fragments, dicentrics and the other centromeric spreading, chromatin bodies, aneuploidy, was carried out for statistical comparison to find

out which specific aberration contributed to the significant difference between experimental group and control. The statistical comparison revealed that chromatin bodies contributed significant difference ($p < 0.001$).

The correlation coefficient 'r' was observed between total chromosomal abnormalities and collection of tissues at various intervals which showed significant correlation ($p < 0.05$) between the two variables in case of liver tissues.

2.2 Interpretation

The study observed an apparent dose-related increase of total chromosomal abnormalities (i.e., breaks, gaps, acentric fragments, dicentrics, rings, centromeric spreading, extreme fragmentation of the chromosomes leading to the formation of chromatin bodies like structures and aneuploidy) in the testicular tissue of 1-day-old pups exposed antenatally to CP in 2/12, 10/12, 20/12; 2/15, 10/15, 20/15 and 2/18, 10/18, 20/18 (doses/gestational age) experimental groups. A similar dose-related increase in chromosome aberrations was observed in liver tissue, whereas the studies conducted in bone marrow from 1-day-old pups antenatally exposed different dosed of CP did not any linear dose–effect relationship. However, the statistical analysis (F-values) of doses and days relationship alone did not show any significant difference for total chromosomal aberrations, suggesting that doses ranging from 2 to 20 at 12–18 days gestation do not lead to any significant difference in terms of induction of chromosomal aberration. But when results obtained at these doses with respect of different gestational ages were compared with control, a significant difference was observed in experiments 10/12, 10/18, 20/12, 20/15, 20/18 in testes and 20/12 in liver and bone marrow, thus indicating that 2 and 10 mg/kg doses of CP caused less genotoxic effect (i.e., less increase in the frequency of chromosomal aberrations), probably due to low metabolic capacity of the target cells, which are incapable of transforming the promutagen into the genotoxic metabolites in heterozygotes [21]. However, a maximum damage was observed in terms of total chromosomal aberration, in all the three tissues studied in 20/12 experimental group, suggesting this gestational age during development to be prone to genotoxicity. Although apparently in case of testicular tissue all the three days of gestations (involving pre- and post-testicular tissue formation periods) used for insulting the developing embryo seemed to be critical for the genotoxic effect, the 12th day of gestation remained most sensitive and critical period for severe chromosomal abnormalities in case of not only the testicular tissue but also liver and bone marrow. This observation was further supported by experiments where single dose of 20 mg/kg of CP was given at 12th day of gestation to pregnant rats and embryonic tissue and/or liver or testes studied at 14th, 16th, 18th and 22nd days of gestation, respectively. It was observed that the total chromosomal aberrations in the embryonic fetal tissues between 14th and 22nd day of gestation showed the following trend: $14 < 16 < 18 < 22$.

A dose-related increase of chromosome aberrations [29] has been reported in preimplantation embryos after maternal administration of CP in spermatogonia, spermatocytes and preovulatory phase in oogonia of mouse [30]. It was observed that in case of spermatogonia, spermatocytes 80 mg/kg dose of CP gave double the amount of chromosome aberrations than at 40 mg/kg. However, this linear dose–effect relationship did not continue to hold true between 80 and 160 mg/kg dose presumably because of the lethal effect upon the sensitive spermatogonia. In in vivo studies in the bone marrow of mouse, rat and Chinese hamster, a similar dose-related increase of chromosome aberrations was observed [25]. Although the present study showed an apparent dose–effect relationship with respect to chromosome aberrations in testes and liver, it was statistically non-significant as observed in bone marrow as well. The reason in this study for not following a dose-related pattern of chromosome aberrations could be various such as the nature of material used which was different from what is reported in literature, the selection of doses, usually in literature dose of CP used are 20 mg/kg and above, or the length of after exposure gap in present experiments. Reports in literature have shown a decrease in chromosome aberration corresponding to the increases in after-exposure gap [22]. It was concluded that CP reached bone marrow and spleen cells of mouse within 6 h of treatment [25] and for the purpose of mutagenicity testing it was recommended to process rat bone marrow 6, 24, 48 h after treatment with a single dose of mutagen [31, 32]. Further, the highest frequency of aberrant cells was indicated to occur for different chemicals during 12–14 h after their single-dose application. It was observed that the frequency of metaphase cells with chromosome aberration produced by, i.e., application of CP differed significantly from these control levels already 6 h after mutagen administration, reaching a maximum between 12 and 18 h. The frequency of aberrant cells was observed to decrease after 24 h [22].

The present study conducted in testicular tissue liver and bone marrow in 1-day-old pups antenatally exposed to CP at 12 or 15 or 18 day of gestation, left an after exposure gap of approximately 10, 7 and 4 days and 4 days in three respective experimental groups. However, a significantly different maximum damage observed in terms of total chromosomal abnormalities in 20/12, experimental groups, not only in testicular tissue but also in liver and bone marrow even after exposure gap of ten days apparently indicated that there was no decrease in chromosomal aberrations even after a prolonged exposure to CP.

Further, it was observed that different categories of chromosomal aberrations, one, including breaks, gaps, acentric fragments, dicentrics, ring chromosomes and the other involving centromeric spreading, chromatin bodies and aneuploid followed two different rules. The former group of chromosomal aberrations decreased with longer after exposure time and the latter group of chromosomal aberrations increased with increasing after exposure. These observations were statistically non-significant when compared to control in the first group of chromosomal aberrations and highly significant in case of the second group of chromosomal aberrations in all the three tissues studies, i.e., testes, liver and bone marrow. These observations further suggested that on one hand, the longer after exposure, gap led to a decrease in aberrations like breaks, gaps, acentric fragments dicentrics and ring chromosomes and on the other hand, increased the aberrations such as centromeric spreading, chromatin bodies and aneuploids. The latter observation is in total contrast with what has been reported earlier in literature. These findings were further confirmed by observing the frequency of above two groups of chromosomal abnormalities in embryonic tissues and/or liver and/or testes of embryo/fetuses examined at various intervals, i.e., 14, 16, 18 and 22 days of gestation after a single-dose exposure of 20 mg/kg CP. A decreasing trend of chromosomal abnormalities such as breaks, gaps, acentric fragments, dicentrics and ring was observed in all the tissues studied from 14th to 22nd day of gestation. The statistical analysis carried out with respect to control in all the tissues also showed no significant difference, whereas in the same experiment chromosomal abnormalities like centromeric spreading, chromatin bodies and aneuploids increased with advancing gestational age. The statistical analysis with respect to control in this case showed a significant difference. It was further observed that chromatin bodies were the one which contributed to this significance. These observations suggested that the yield of one group of the chromosomal aberrations by the length of the exposure to CP. In the former situation, the length of the exposure whereas in the latter the after-exposure interval did not have any bearing on the yield of chromosomal aberrations.

Interestingly, a highly significant difference in term of chromosomal abnormalities such as centromeric spreading, chromatin bodies and aneuploidy was observed in the liver tissue of normal and abnormal looking fetuses collected at 22nd day of gestation, thus indicating that chromosomal aberrations might be an expression of a direct cytotoxic effect [33]. The presence of fragmented chromosomes leading to chromatin bodies probably indicated that these aberrations were elicited in cell at G-phase of the cell cycle and the reasons to find a highly significant correlation of this type of chromosomal damage at a particular dose (20 mg/kg) of CP in all the tissues studied could as well be because of the growing embryonic/fetal tissue. It has been observed previously that the sensitivity of regenerating liver cells to CP, in comparison to the normal liver cells (of non-hepatectomized rats) and bone marrow was more in including chromosomal aberration [21]. It is suggested that the study of 'chromatin bodies' in developing embryo/fetuses could be as a potential index of genotoxicity.

The results in the present study indicated that centromere spreading and aneuploidy contributed to the significant difference when results from 10/12 and 10/18 experimental groups were compared with control. A maximum frequency of aneuploidy was, however, observed in testicular tissues of 1-day-old pups exposed antenatally to 10 mg/kg dose of CP and at 12th day of gestation. It was further observed that aneuploidy in C- and D-group chromosomes contributed to this high frequency. A similar observation with respect to both C- and D-group chromosomes was made in 20/15 experimental group. The correlation between the frequency distribution of aneuploidy of each chromosomes within C- and D-group chromosomes (i.e., aneuploidy of 11–13 in C-group and 14–20 in D-group per metaphase plate) showed a significant difference within D-group chromosomes in 10/12, 10/15 and 10/18 experimental groups, suggesting that CP probably leads to a non-random, non-disjunction event in C- and D-group chromosomes. These observations are in contrast to what has been observed previously where they indicated that 30, 50, 145 mg/kg dose of CP were capable of inducing non-disjunction. The possible reasons for aneuploidy, however, could be either as a consequence of chromosome breakage and rearrangement or as a secondary event to clastogenesis. The present study showed that at a particular dose level, single acute treatment with CP was capable of inducing non-disjunction events in developing testicular tissue. The qualitative discrepancies with respects to dose in literature and the present study could as well be due to the difference in the physiology of cell kinetics in two different test systems (materials) used.

References

1. P.E. Polani, DNA repair defects and chromosome instability disorders. Ciba Found. Symp. **66**, 81–131 (1979)
2. F. Mitelman, Catalogue of chromosome aberrations in cancer. Cytogenet. Cell Genet. **36**(1–2), 1–515 (1983)
3. R. Albanese, Sodium fluoride and chromosome damage (in vitro human lymphocyte and in vivo micronucleus assays). Mutagenesis **2**(6), 497–499 (1987)
4. D.A. Preston, R.N. Jones, A.L. Barry, C. Thornsberry, Comparison of the antibacterial spectra of cephalexin and cefaclor with those of cephalothin and newer cephalosporins: reevaluation of the class representative concept of susceptibility testing. J. Clin. Microbiol. **17**(6), 1156–1158 (1983)
5. I. Kodýtková, J. Madar, R.J. Srám, Chromosomal aberrations in mouse bone marrow cells and antibody production changes induced by long-term exposure to cyclophosphamide and alpha-tocopherol. Folia Biol. (Praha) **26**(2), 94–102 (1980)
6. N.P. Bochkov, R.J. Sram, N.P. Kuleshov, V.S. Zhurkov, System for the evaluation of the risk from chemical mutagens for man: basic principles and practical recommendations. Mutat. Res. **38**(3), 191–202 (1976)
7. G.R. Mohn, J. Ellenberger, Genetic effects of cyclophosphamide, ifosfamide and trofosfamide. Mutat. Res. **32**(3–4), 331–360 (1976)
8. O.P. Kirichenko, A.N. Chebotarev, Relationship between the cytogenetic effect of different concentrations of thiophosphamide and phosphemide and the duration of their contact with human lymphocytes. Genetika **12**(6), 142–149 (1976)
9. P.E. Crossen, Variation in the sensitivity of human lymphocytes to DNA-damaging agents measured by sister chromatid exchange frequency. Hum. Genet. **60**(1), 19–23 (1982)
10. E. Schmid, M. Bauchinger, Inter- and intrachromosomal distribution of cyclophosphamid-induced chromosome aberrations in man. Mutat. Res. **9**(4), 417–424 (1970)
11. J.W. Allen, S.A. Latt, Analysis of sister chromatid exchange formation in vivo in mouse spermatogonia as a new test system for environmental mutagens. Nature **260**(5550), 449–451 (1976)
12. W. Vogel, T. Bauknecht, Differential chromatid staining by in vivo treatment as a mutagenicity test system. Nature **260**(5550), 448–449 (1976)
13. T. Bauknecht, W. Vogel, U. Bayer, D. Wild, Comparative in vivo mutagenicity testing by SCE and micronucleus induction in mouse bone marrow. Hum. Genet. **35**(3), 299–307 (1977)
14. E.L. Schneider, Y. Nakanishi, J. Lewis, H. Sternberg, Simultaneous examination of sister chromatid exchanges and cell replication kinetics in tumor and normal cells in vivo. Cancer Res. **41**(12 Pt 1), 4973–4975 (1981)
15. A. Korte, G. Obe, I. Ingwersen, G. Rückert, Influence of chronic ethanol uptake and acute acetaldehyde treatment on the chromosomes of bone-marrow cells and peripheral lymphocytes of Chinese hamsters. Mutat. Res. **88**(4), 389–395 (1981)

16. H. Nau, H. Spielmann, C.M. Lo Turco Mortler, K. Winckler, L. Riedel, G. Obe, Mutagenic, teratogenic and pharmacokinetic properties of cyclophosphamide and some of its deuterated derivatives. Mutat. Res. **95**(2–3), 105–118 (1982)
17. F. Darroudi, A.T. Natarajan, Cytological characterization of Chinese hamster ovary X-ray-sensitive mutant cells, xrs 5 and xrs 6. II. Induction of sister-chromatid exchanges and chromosomal aberrations by X-rays and UV-irradiation and their modulation by inhibitors of poly(ADP-ribose) synthetase and alpha-polymerase. Mutat. Res. **177**(1), 149–160 (1987)
18. A. Manyak, E. Schleiermacher, Action of mitomycin C on mouse spermatogonia. Mutat. Res. **19**(1), 99–108 (1973)
19. C.G. Goetz, Defining drug-induced supersensitivity. Am. J. Psychiatry **137**(8), 992–993 (1980)
20. J. Ashby, B. Beije, Concomitant observations of UDS in the liver and micronuclei in the bone marrow of rats exposed to cyclophosphamide or 2-acetylaminofluorene. Mutat. Res. **150**(1–2), 383–392 (1985)
21. F. Spirito, M. Rizzoni, E. Lolli, C. Rossi, Reduction of neutral gene flow due to the partial sterility of heterozygotes for a linked chromosome mutation. Theor. Popul. Biol. **31**(2), 323–338 (1987)
22. R.J. Srám, V.S. Zhurkov, J. Nováková, I. Kodýtková, Changes in the frequency of chromosome aberrations in the bone marrow of mice examined at various intervals after single-dose and continual exposures to cyclophosphamide. Folia Biol. (Praha) **27**(1), 58–65 (1981)
23. Y. Nakanishi, D. Kram, E.L. Schneider, Aging and sister chromatid exchange. IV. Reduced frequencies of mutagen-induced sister chromatid exchanges in vivo in mouse bone marrow cells with aging. Cytogenet. Cell Genet. **24**(1), 61–67 (1979)
24. G. Krishna, J. Nath, T. Ong, W.Z. Whong, A simple method for the extraction of mutagens from airborne particles. Environ. Monit. Assess. **5**(4), 393–398 (1985)
25. G. Krishna, J. Nath, M. Petersen, T. Ong, Cyclophosphamide-induced cytogenetic effects in mouse bone marrow and spleen cells in in vivo and in vivo/in vitro assays. Teratog. Carcinog. Mutagen. **7**(2), 183–195 (1987)
26. P.K. Datta, E. Schleiermacher, The effects of cytoxan on the chromosomes of mouse bone marrow. Mutat. Res. **8**(3), 623–628 (1969)
27. E. Schleiermacher, W. Schmidt, Changes of the synaptonemal complex at the end of pachytene. Humangenetik **19**(3), 235–245 (1973)
28. A.M. Malashenko, N.I. Surkova, Mutagenic effect of thioTEPA in laboratory mice. V. Influence of female genotype on realization of dominant lethal mutations induced in the spermatids of males. Genetika **11**(1), 105–111 (1975)
29. I. Kola, P.I. Folb, An assessment of the effects of cyclophosphamide and sodium valproate on the viability of preimplantation mouse embryos using the fluorescein diacetate test. Teratog. Carcinog. Mutagen. **6**(1), 23–31 (1986)
30. G. Röhrborn, O. Kühn, I. Hansmann, K. Thon, Induced chromosome aberrations in early embryogenesis of mice. Humangenetik **11**(4), 316–322 (1971)
31. N.P. Bochkov, T.V. Filippova, S.M. Kuzin, S.V. Stukalov, Cytogenetic effects of cyclophosphamide on human lymphocytes in vivo and in vitro. Mutat. Res. **159**(1–2), 103–107 (1986)
32. W.W. Nichols, Mutations as environmental hazards. Lakartidningen **69**(23), 2785–2788 (1972)
33. H.G. Payne, L.W. Nelson, J.H. Weikel Jr., Effects of cyclophosphamide on somatic cell chromosomes in rats. Toxicol. Appl. Pharmacol. **30**(3), 360–368 (1974)

Chapter 3
Effect of Cyclophosphamide on Testis—Protein Profile Study

3.1 Introduction

The protein is an important aspect of the process of differentiation and morphogenesis [1]. Since then most of the studies have been histochemical in nature describing the nucleohistones and the localization of the testicular enzymes [2].

Some investigators have analyzed changes which occur during development and during the maturation of rat testes. Mills et al. (1972) reported that total DNA, RNA and protein contents decreased in testicular tissue during development and corresponded to cell population and morphological changes. They observed that the protein content of the rat testis represented an average of 66% of the total testis dry weight. The protein content of the testis per gram weight decreased as the dry weight decreased during the development. The decreased in percentage dry weight of the testes corresponded closely with the lumen formation in the seminiferous tubules. Further, gonadal patterns in rat during development from the morphologically indifferent stage up to birth were studied by two-dimensional gel electrophoresis by Muller et al. [3]. Specific proteins were detected in both the male and the female sex at the morphologically indifferent stage and during differentiation.

Most of the information regarding protein profile changes in rodents, however, is available in matured gonad both in vivo and in vitro conditions, especially at the time of spermatogenesis. A large number of studies have been performed in order to understand morphological changes which occur during completion of the spermatogenic process [4]. A unique acid-soluble basic protein was detected in the testes of sexually mature rats and a number of other mammalian species. This protein was devoid of phenylamine, tryptophan, glutamic acid, glutamine, isoleucine and cysteine [5]. However, majority of studies concerned with biochemical changes have attempted to correlate enzyme activity to particular cell types in germinal epithelium [6–8]. The expression of many testicular enzymes and proteins is temporarily regulated during spermatogenesis [9]. The synthesis of lactate dehydrogenase-X, a testes-specific isoenzyme, has been shown to begin during the mid-pachytene stage of spermatocyte

© The Author(s), under exclusive license to Springer Nature Singapore Pte Ltd. 2020
A. K. Saxena and A. Kumar, *Fish Analysis for Drug and Chemicals Mediated Cellular Toxicity*, SpringerBriefs in Applied Sciences and Technology,
https://doi.org/10.1007/978-981-15-4700-3_3

development and continue throughout spermatid differentiation [10]. Another testes-specific isoenzyme detected in pachytene spermatocytes but increases markedly in activity as spermiogenesis proceeds [11] analyzed stage-specific protein synthesis during spermatogenesis. Stern (1983) observed quantitative changes for polypeptides of molecular weight ranges from 16,500 to 82,000 Da during spermatogenesis.

However, little is known about the testicular and extratesticular origin of newly synthesized proteins accumulated in each testicular compartment. Cultured rat sertoli cells synthesize and secrete proteins that are serum like [12, 13], follicle stimulating hormone (FSH)-dependent and cell-specific [14, 15]. In seminiferous tubules of the rat, the sertoli cell is influenced cyclically by the spermatogenic cycle [16] reported three acidic proteins found in testicular intertubular fluid (TIF) and seminiferous tubular fluid (TIF) of molecular weight 72,000, 45,000 and 35,000 Da possibly secreted by sertoli cells. They suggested that most albumin and transferrin found in TIF and SNF have an extratesticular origin and that proteins secreted by the Sertili cells could gain access to both TIF and SNF. It is not surprising to find sertoli cell-specific secretory proteins in TIF samples since androgen-binding protein, another sertoli cell-specific secretory protein, has been reported to be secreted into TIF [17].

Also, the role of retinoids in mammalian spermatogenesis has long been recognized. Wolbach and Howe [18] reported the failure of spermatogenesis in animals deprived of retinol. Later studies [19] showed that in retinol deficiency, spermatogenesis does not progress beyond early meiosis and is accompanied by extensive degeneration of the germinal epithelium. Retinol is transported in plasma bound to a specific transport protein, retinol-binding protein [20]. The cellular localization of the different retinoid binding proteins in the rat testes has been examined in several reports [21, 22]. The cytosolic retinol-binding protein (CRBF) has been found to be strikingly localized within the sertoli cells [23]. In contrast, cellular retinoic acid-binding protein (CRABP) was found particularly in germ cells. It has also shown enrichment of CRBP in sertoli cells and that of CRABP in germ cells.

From both an applied and basic standpoint, a study of antineoplastic drug such as cyclophosphamide on developing testes is valuable. The present study was initiated with the aim to learn the about the action of CP on developing male gonad (testes) in terms of protein profile changes, after insulting the developing fetuses with different doses of CP.

Procedure: Animals: Three gestational ages (12th, 15th and 18th) with three doses (2, 10, 20 mg/kg) of CP have already been described earlier.

Chemicals: The chemicals used were acrylamide (Singma Chem. Co., USA); N,N'-methylene-bis-acrylamide(Bis); N,N,N',N'-tetramethylethylenedimamine (TEMED, Sigma Chem. Co., USA); sodium dodecyl sulfate (SDS, Sigma Chem. Co., USA); phenylmethyl sulfonyl fluoride (PMSF, Sigma Chem. Co., USA); ammonium persulfate (Sigma Chem. CO.'USA); Coomassie Brilliant Blue R-250 Sigma Chem. CO.' USA); silver nitrate (BDH, India); potassium dichromate (BDH, India); sodium carbonate (BDH, India); potassium tartarate (BDH, India); Folin's Reagent (BDH, India); bovine serum albumin (Sigma Chem. Co., USA); copper sulfate (BDH, India); ß-mercaptoethanol (BDH, India); Triton X-100 (Centronic, India); magnesium chloride (BDH, India); sodium metabisulfite (BDH, India);

Bromophenol Blue (Dye) (E. Merck's England); potassium chloride (BDH, India); nitric acid (BDH, India); formaldehyde (BDH, India); hydrochloric acid (BDH, India); methyl alcohol (BDH, India); ethyl alcohol (absolute) (BDH, India); glycerol (BDH, India); acetic acid (BDH, India); glycine (BDH, India); trichloroacetic acid (TCA) (BDH, India).

Preparation of Homogenizing Buffer

Tris (hydroxylmethylaminomethane)	10.00 mM
Potassium chloride	9.00 mM
Magnesium chloride	0.250 mM
Sodium metabisulfite	0.025 mM
ß-mercaptoethanol	6.000 mM
Phenylmethyl sulfonyl fluoride	0.025 mM
Triton X-100	2%

The pH of the homogenizing buffer was adjusted 7.4 with HCL.

Preparation of Electrophoresis Buffer

Tris (hydroxymethylaminomethane)	0.625 M
Glycine	0.192 M
Sodium dodecyl sulfate	0.01%

The pH of the electrode buffer was adjusted 8.3 with the hydrochloride acid.

Preparation of Stacking Gel Buffer

Tris (hydroxymethylaminomethane)	0.125 M
Sodium dodecyl sulfate	0.1%

The pH of the stacking buffer was adjusted to 6.8 with hydrochloride acid.

Preparation of Separating Gel Buffer

Tris (hydroxymethylaminomethane)	0.375 M
Sodium dodecyl sulfate	0.1%

The pH of the separating buffer was adjusted to 8.8 with hydrochloride acid.

Preparation of Sample Buffer

Tris (hydroxymethylaminomethane)	0.0625 M
Sodium dodecyl sulfate (SDS)	4%
Glycerol	10%
ß-mercaptoethanol	5%
Bromophenol Blue (Dye)	0.001%

The pH of the sample buffer was adjusted to 6.8 with the help of hydrochloride acid.

Preparation of Gel Solution

Stock solution of 30% by weight of acrylamide and 0.8% by weight of N-N'-bismethylene acrylamide was prepared. The stacking gel 2.5% was prepared by taking 2.5 ml of stock solution and 10 and 15% separating gels prepared by taking 10 and 15 ml of the stock solution. The gels were polymerized chemically by the addition of 0.025% by volume of N, N, N, N-tetramethylethylenediamine (TEMED) and ammonium persulfate. After mixing, the gel solution was immediately placed in glass and allowed to stand for 20–30 min for polymerization. Just before the use, the water layer was sucked off and gel was placed in an electrophoresis apparatus.

Preparation of Protein Reagent

This reagent was freshly prepared before use. 100 ml pf 2% Na_2CO_3 in 0.1 NaOH was taken and to it 1 ml of 4% potassium tartarate added. After shaking well, this mixture 1.0 ml of 2% $CuSO_4$ was added drop by drop followed by vigorous shaking to avoid turbidity.

Procedure

Testicular tissue was collected from one-day-old pups in nine parallel sets of experiment where three different doses—2 or 10 or 20 mg/kg body weight of CP were administered intraperitoneally to the pregnant rats at any of the three different gestational ages, i.e., 12th or 15th or 18th day. These experiments were designated according to dose/gestation as 2/12, 10/12, 20/12; 2/15, 10/15, 20/15; and 2/18, 10/18, 20/18 as documented in Table 3.1, which also depicts the number of mother rats used and total number of pups studied.

   Some of the pups from these experimental groups were allowed to grow for three to four months as adults and mated with respective opposite's sexes to find out if infertility was induced.

   In another set of experiment, 5 mg/kg body weight of CP was injected (i.p.) to pregnant rats during 12th to 15th day of gestation and male litters examined for infertility at three to four months of age (examined by mating with respective controls). Testicular tissue from such infertile males was also subjected to the study of protein profile. Testes from 1-day-old pups of normal rats and that from the age matched normal male adult rats were used as controls.

   In a third set of experiment (as proposed in the aims and objectives), the extract of the middle portion of the male embryos (excluding the anterior and posterior portions with fore and hind limbs of the embryo, only to include the region-bearing the testicular tissue) was used for protein profile studies at 12th, 13th, 14th and 15th day of gestation. The rationale for doing these experiments was to know if the proteins, which normally appeared in 12th to 15th day embryos, were affected by CP given to mother rats on 12th or 15th day. The embryo in this case was sexes on the basis of the presence of the absence of the X-chromatin body in the interphase cells and sometimes on the basis of the karyotypes.

Table 3.1 Number of 1-day-old male pups exposed antenatally to CP used for protein profile study

Gestational period (days)	Doses of CP (mg/kg)	Experimental code	Number of mother rats used	Number of male pups delivered	Number of female pups delivered	Number of male pups used for biochemical studies
Control	–	–	7 (13)	18 (25)	13 (32)	6
12	2	2/12	4 (11)	8 (29)	9 (27)	3
	10	10/12	5 (13)	13 (33)	14 (33)	3
	20	20/12	6 (15)	11 (35)	16 (34)	10
15	2	2/15	5 (13)	13 (42)	4 (28)	4
	10	10/15	3 (11)	12 (32)	13 (26)	4
	20	20/15	4 (11)	11 (25)	9 (24)	4
18	2	2/18	4 (10)	15 (36)	17 (40)	5
	10	10/18	5 (12)	8 (24)	20 (44)	4
	20	20/18	6 (13)	17 (31)	17 (28)	5

CP Cyclophosphamide
() Data in the parenthesis indicates the total number studied for biochemical study

Processing of the Tissues

The fresh tissue was homogenized in the ratio of 100 mg/2.5 ml (w/v) at 4 °C for 10 min in the homogenizing buffer (pH 7.4). The homogenate was centrifuged at 4000 × g for 30 min at 4 °C. The supernatant and the pellet were saved for electrophoresis. The protein concentration was estimated by Lowry's method as tabulated below, using crystalline BSA as standard.

Quantification of protein by Lowry's method

Reagent	Blank	1	2	3	4	5
Water	0.5	0.4	0.3	0.3	0.2	–
Protein solution	–	0.1	0.2	0.2	0.3	0.5
Protein reagent	5.0	5.0	5.0	5.0	5.0	5.0

Mixed and allowed to stand for exactly 10 min and then added.

Folin's Reagent 0.5 0.5 0.5 0.5 0.5 0.5

Mixed and allowed to stain for another 10 min. and absorbance measures at 660 against the blank.

The standard graph was drawn plotting optical density against concentration of known standard protein. The amount of protein in the samples was determined. The purpose of determining the amount of protein in a given sample was to ensure loading of equal amounts of protein in gels in all the experimental groups for a precise quantitative comparison.

Running of Gels

The proteins were run on 10 and 15% SDS-polyacrylamide slab gel containing 0.1% SDS in final concentration. The discontinuous buffer system of Laemmli (1970) was used to separate proteins according to their molecular weight. The protein samples were completely dissociated in sample buffer, keeping in boiling water for 10 min in each slot, samples containing 50 μg m of protein were loaded and run at constant current 100 v.

Staining of Gels with Coomassie Blue and Silver

After the run, the gels were fixed in 50% TCA overnight, washed with water and then stained with freshly prepared 0.1% staining solution of Coomassie Brilliant Blue R-250. The staining solution was prepared in a mixture of methanol: acetic acid: water in the ratio of 5:1:5. The gels were photographed when maximum intensity of bands appeared and further destained for photosensitive silver staining by the method of Merrill's (1981). The gels were treated through a series of steps as follows:

1. 50% methanol + 12% acetic acid—20 min
2. 10% ethanol + 5% acetic acid—10 min (three times, 200 ml)
3. 0.003M Potassium dichromate + 0.0032N nitric acid (200 ml)
4. Distilled water 30 s—four times (200 ml)
5. 0.012 M AgNO$_3$—200 ml—30 min.
6. Rinsing in 0.28M sodium carbonate +0.5 ml of formaline per liter (300 ml)—three times (first two rinsing with developer then develop in third 300 ml of developer).
7. Developing was stopped by discarding the developer and adding 1% acetic acid (100 ml).
8. Gels were washed twice with 200 ml of distilled water and stored in 1% acetic acid solution.

The double staining technique was used for visualization and comparison of protein bands in the same SDS-polyacrylamide gel by the method of Saxena and Bamezai [24]. The synopsis of the sequences used in destaining the Coomasie blue gel and further staining with silver staining are documented as follows:

Steps	Solutions	Times
1.	Gels destained in 50% methanol; 7% acetic acid; 200 ml four times; 6-h interval	24 h
2.	Wash with distilled water (200 ml)	1–2 min
3.	5% HCl (200 ml—two times)	5 min
4.	Wash with distilled water—200 ml four times—30 min each	2–4 h
5.	Gels fixed in 50% methanol: 10% acetic acid	20 min–12 h

(continued)

(continued)

Steps	Solutions	Times
6.	Silver stained with little modification of Merrill's method (1081)	
7.	Before developing rinsed with 5% formalin	1–2 min
8.	10% acetic acid (200 ml)	
9.	Wash with distilled water and photographed	

Molecular Weight Determination

The molecular weight of the polypeptides was determined by the method of Weber and Osborn [25]. The protein standards covering a sufficiently broad molecular weight range were used to construct standard plots of log molecular weight versus mobility for each of the polyacrylamide gel concentration used. The following protein standards provided an acceptable molecular weight range: bovine serum albumin (BSA); (MW 6600); egg bovine (MW 45,000); glyceraldehye 3-phosphate dehydrogenase (G-3-P D) (MW 36,000); carbonic anhydrase (MW 29,000); trypsinogen (MW 24,000); trypsin (MW 20,100); α-Lacta albumin (MW 14,200). The mobility was calculated as:

Distance of protein migration Length before staining

$$\text{Mobility} = \frac{\text{Distance of protein migration}}{\text{Length after destaining}} \times \frac{\text{Length before staining}}{\text{Distance of dye migration}}$$

The relative mobilities were plotted against the known molecular weights to construct standard graph of log molecular weight.

The standard graph—log known molecular weight vs. mobility for 10 and 15% SDS-polyacrylamide gel.

Findings

The protein profile changes were observed in developing testes, collected from 1-day-old pups, antenatally to a single dose (i.e., 2 or 10 or 20 mg/kg body weight) of CP during 12- or 15- or 18-day gestation. The qualitative changes were observed in both 10 and 15% SDS-polyacrylamide gels after using double-staining technique. The detailed and pooled data of both the gels is mentioned in tables.

Table 3.2 presents the qualitative changes in protein bands observed in the testicular tissue of 1-day-old pups exposed antenatally to CP at 12th day of gestation with respect to age matched control. The most prominent protein profile changes in supernatant fraction were as follows: (a) the disappearance of intense protein band at approximately 4400 Da in 2/12 experimental groups, apart from the disappearance of less than 20,000 Da protein band in the intense protein band at approximately 86,000, 74,500 and 64,000 Da in 2/12 experimental groups; 86,000 and 74,000 Da in 10/12 and 20/12 experimental groups; and (c) the increased intensity of protein bands of approximately less than 20,000 Da in 10/12 experimental groups. The pellet fraction also showed profile changes which were: (a) the disappearance of protein

Table 3.2 Qualitative protein profile changes of testes from 1-day-old pups exposed antenatally to single dose of CP at 12th day of gestation

Supernatant fraction				Pellet fraction			
Protein bands in 1-day-old control testes (Da)	Qualitative changes in protein bands with respect to control at different doses CP (mg/kg)			Protein bands in 1-day-old control testes (Da)	Qualitative changes in protein bands with respect to control at different doses CP (mg/kg)		
	2	10	20		2	10	20
86,000	DEC	DEC	DEC	74,500	DEC	DEC	INC
74,000	DEC	DEC	DEC	64,000	DEC	DIS	NA
64,000	DEC	DIS	DIS	58,000	DEC	DIS	NA
58,000	NA	DIS	DIS	44,000	NA	DEC	INC
44,000	DIS	DIS	DIS	24,000	DIS	DEC	INC
20,000	NA	INC	DIS	21,000	NA	INC	INC

Not affected; *DEC* decreased; *INC* increased; *DIS* disappeared. Pooled data from 10 and 15% SDS-PAG

band at approximately, 24,000 Da in 2/12 experimental group; 6400 and 5800 Da in 10/12 experimental group; (b) the reduction of the intensity of protein bands at approximately 74,500, 64,000, 58,000 Da in 2/12 experimental group and 74,500, 44,000, 24,000 and 21,000 Da in 2/12 experimental groups and (c) the increase in the intensity of protein band at 20/12 experimental groups.

The qualitative changes in protein profile in testicular tissue from 1-day-old pups exposed antenatally to CP at 15th day of gestation are depicted in Table 3.3. The most

Table 3.3 Qualitative protein profile changes of testes from 1-day-old pups exposed antenatally to single dose of CP at 15th day of gestation

Supernatant fraction				Pellet fraction			
Protein bands in 1-day-old control testes (Da)	Qualitative changes in protein bands with respect to control at different doses CP (mg/kg)			Protein bands in 1-day-old control testes (Da)	Qualitative changes in protein bands with respect to control at different doses CP (mg/kg)		
	2	10	20		2	10	20
86,000	NA	NA	INC	74,500	NA	DEC	NA
74,000	NA	NA	INC	64,000	NA	DIS	DEC
64,000	NA	NA	INC	58,000	NA	DEC	DEC
58,000	NA	NA	INC	44,000	NA	DEC	DEC
44,000	NA	DIS	INC	24,000	NA	DIS	DIS
20,000	NA	NA	INC	21,000	NA	DIS	DIS

NA Not affected; *DEC* decreased; *INC* increased; *DIS* disappeared
Pooled data from 10 and 15% SDS-PAG

Table 3.4 Qualitative protein profile changes of testes from 1-day-old pups exposed antenatally to single dose of CP at 18th day of gestation

Supernatant fraction				Pellet fraction			
Protein bands in 1-day-old control testes (Da)	Qualitative changes in protein bands with respect to control at different doses CP (mg/kg)			Protein bands in 1-day-old control testes (Da)	Qualitative changes in protein bands with respect to control at different doses CP (mg/kg)		
	2	10	20		2	10	20
86,000	NA	DEC	DEC	74,500	NA	NA	DIS
74,000	NA	DEC	DEC	64,000	NA	NA	DIS
64,000	NA	DEC	DEC	58,000	NA	NA	NA
58,000	NA	DEC	DEC	44,000	NA	DEC	NA
44,000	NA	DEC	NA	24,000	NA	DEC	INC
20,000	NA	NA	INC	21,000	NA	DEC	INC

NA Not affected; *DEC* decreased; *INC* increased; *DIS* disappeared. Pooled data from 10 and 15% SDS-PAG

prominent changes in supernatant fraction were as follows: (a) the disappearance of intense protein band at approximately 44,000 Da in 10/15 experimental group and (b) the increased intensity of protein bands at approximately 86,000, 64,000, 58,000, 44,000 and less 20,000 Da in 20/15 experimental groups.

The following protein profile changes were observed in pellet fraction: (a) the disappearance of intense protein band at approximately 64,000, 24,000, 21,000 Da in 10/15 experimental group; 24,000 and 12,000 Da in 20/15 experimental group; (b) reduction of the intensity of intense protein bands at approximately 74,500, 58,000, 44,000 Da in 10/15 experimental group and 64,000, 58,000 and 44,000 Da in 20/15 experimental groups.

The protein profile changes in testicular tissue of 1-day-old pups exposed antenatally to CP on 18th day of gestation are shown in Table 3.4. The most prominent changes occurred in supernatant fraction which were as follows: (a) the disappearance of decrease in the intensity of intense protein bands at approximately 86,000, 74,000 and 64,000, 58,000 Da in 20/18 experimental group and 86,000, 74,500, 64,000, 58,000, 44,000 Da in 10/18 experimental group and (b) the increase intensity of protein band at approximately less than 20,000 Da observed in 20/18 experimental group. The pellet fraction also showed qualitative changes in testicular tissue: (a) the disappearance of intense protein bands at approximately 75,000 and 64,000 Da in 20/18 experimental group and (b) the reduction of intensity of protein band at approximately 24,000 and 21,000 Da in 20/18 experimental groups.

The experimental plan also showing the collection of tissue from normal developing embryos of male or female pups for protein profile study at 12th to 15th day of gestation to find out stage or sex-specific proteins in both supernatant and pellet fractions were analyzed by SDS-polyacrylamide gel electrophoresis to know if the affected protein profile in the previous experiments involved any of the proteins which appear during these gestational ages.

The study of 12th-day male embryo supernatant fractions showed protein bands at approximately 135,000, 127,000, 125,000, 118,000, 110,000, 104,000, 96,000, 58,000, 44,000, 24,000 and three bands of approximately less than 20,000 Da. A similar pattern of protein profile was observed in female embryos, of the same gestational age, except for the disappearance of one protein bands of approximately less than 20,000 Da (Table 3.5).

Male embryo of 13th day gestational age showed the appearance of protein band of approximately 135,000, 127,000, 125,000, 118,000, 110,000, 104,000, 96,000, 66,000, 58,000, 44,000, 24,000 Da and three protein bands of less than 20,000 Da in the supernatant fraction, whereas the female embryo of same gestational age

Table 3.5 Protein profile changes in the supernatant fraction of normal developing embryos

Mol. Wt. (Da)	Gestational age (days)							
	12		13		14		15	
	M	F	M	F	M	F	M	F
146,000	–	–	–	–	–	–	–	+
142,000	–	–	–	–	+	–	+	+
136,000	–	–	–	–	+	+	+	+
135,000	+	+	+	–	–	–	–	–
130,000	–	–	–	–	–	–	+	+
127,000	+	+	+	–	–	–	+	+
125,000	+	+	+	–	–	–	–	–
121,000	–	–	–	–	+	+	+	+
118,000	+	+	+	+	–	–	–	–
117,000	–	+	–	–	+	+	+	+
110,000	+	–	+	+	+	+	+	+
96,000	+	–	–	–	+	+	+	+
67,000	–	+	+	–	–	–	–	–
66,000	–	–	+	–	–	–	–	–
58,000	+	+	–	–	+	+	+	+
52,000	–	–	+	+	+	+	+	+
44,000	+	+	–	–	+	+	+	+
37,000	–	–	+	–	+	+	+	+
24,000	+	+	+	+	+	+	+	+
20,000	+	–	–	+	+	+	+	+
20,000	+	+	+	+	+	+	+	+
20,000	+	+	+	+	–	–	+	+

M Male; *F* female; (+) present; (–) not present; data from 10 and 15% SDS-PAG showed appearance of protein bands of approximately 118,000, 110,000, 104,000, 96,000, 66,000, 44,000, 24,000 and two bands of less 20,000 Da. Table 3.5 shows the protein profile in 14th day male embryo supernatant fraction where the protein bands at approximately 142,000, 136,000, 121,000, 117,000, 110,000, 71,000, 67,000, 52,000, 44,000, 37,000 and two bands of less than 20,000 Da were observed. The female embryo showed the same pattern except for the disappearance of 142,000 Da protein band.

The protein profile of 15th embryo showed the appearance of protein bands of approximately 142,000, 136,000, 130,000, 127,000, 121,000, 117,000, 110,000, 71,000, 67,000, 52,000, 44,000, 37,000 and three bands of approximately less than 20,000 Da in the supernatant fraction, whereas female embryos of the same gestation showed similar pattern of protein profile except for the presence of 146,000 Da protein band (Table 3.5). The protein profile changes in supernatant and pellet fraction of 1-day-old testes exposed antenatally at 12th, 15th, 18th days of gestation with different doses of CP, using 10 to 15% SDS-PAGE stained with Coomassie blue and the same gel for silver staining were used for protein profile study and comparison of protein bands using two different staining in the same PAGE (Fig. 3.1).

Table 3.6 depicted the protein profile observed from the pellet fraction of 12th, 13th, 14th, 15th day embryo using 15% gel. The 12th day male embryos showed the presence of protein bands of approximately 92,000, 82,000, 58,000, 42,000, 38,000, 32,000, 30,000 and two bands of less than 20,000 Da, whereas the female of the same

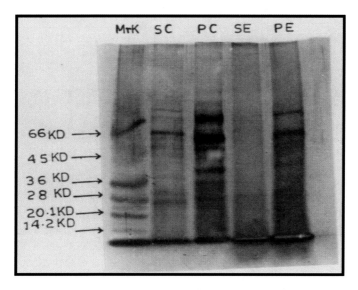

Fig. 3.1 Protein profile changes in supernatant and pellet fraction of the normal developing embryonic tissue collected from lower middle region of abdomen containing gonadal ridge from 12th to 15th day of gestation, using 15% SDS-PAG with silver staining. (Mrk = Marker; SC = Supernatant control; PC = Pellet control; SE = Supernatant experiment; PE = Pellet experiment)

Table 3.6 Protein profile changes in the pellet fraction of normal developing embryos

Mol. Wt. (Da)	Gestational age (Days)							
	12		13		14		15	
	M	F	M	F	M	F	M	F
94,000	–	–	–	–	+	+	+	+
92,000	+	+	+	+	–	–	–	
84,000	–	–	–	–	–	+	–	–
82,000	+	+	+	+	+	+	+	+
78,000	–	–	–	–	+	+	+	+
68,000	–	–	–	–	+	+	+	+
58,000	+	+	+	+	–	–	–	–
55,000	–	–	–	–	+	+	–	–
50,000	–	–	–	–	–	–	+	–
48,000	–	–	+	–	–	–	–	–
46,000	–	–	–	–	–	+	–	–
38,000	+	+	+	+	–	–	–	–
35,000	–	–	–	–	+	+	+	+
32,000	+	–	+	–	–	–	–	–
30,000	+	+	+	–	–	–	–	–
27,000	–	–	–	–	+	+	+	+
22,000	–	–	–	–	+	+	+	+
20,000	–	–	+	+	+	+	+	+
20,000	+	+	+	+	+	+	+	+
20,000	+	+	+	+	+	+	+	+

M Male; *F* female; (+) present; (–) not present; data from 10 and 15% SDS-PAG

gestational age showed a similar pattern except for the disappearance of 32,000 Da protein band.

Male embryos of 13th day gestation showed the appearance of protein bands at approximately 92,000, 82,000, 58,000, 48,000, 42,000, 38,000, 32,000, 30,000 and four band of approximately less than 20,000 Da. The female embryos of the same gestational age showed a similar pattern, except for the disappearance of protein band approximately 48,000, 32,000 and 30,000 Da. In case of day 14th male embryos, the appearance of protein bands of approximately 94,000, 82,000, 78,000, 68,000, 55,000, 42,000, 35,000, 27,000, 22,000, and four bands less than 20,000 Da were observed. The female embryos of same gestation showed a similar pattern of protein profile except for the appearance of two proteins of approximately 84,000 and 46,000.

The 15th day embryos showed the appearance of protein bands of approximately 94,000, 82,000, 78,000, 68,000, 50,000, 42,000, 35,000, 27,000, 22,000, and four bands less than 20,000 Da, whereas the female embryo of the same gestational age group showed similar pattern of protein profile except for the disappearance of approximately 50,000 Da.

Interestingly, it was observed that the protein of 58,000, 44,000, 24,000, and less than 20,000 Da affected in 2/12, 10/12; 20/12; 10/15; 20/15 and 10/18, 20/18 experimental groups; appeared during 12–15 days of gestation. However, it was not studied earlier if these were same proteins required for normal differentiation and further development of male gonad (testis).

Protein Profile Changes in Infertile Testes

The nine parallel sets of experiments where three different doses of CP 2 or 10 or 20 mg/kg body weight were given to the mother rats at 12th or 15th or 18th day of gestation showed no effect on fertility except for the delay in males. However, the infertility was induced where continuous dose of 5 mg/kg of CP was given at 12–15 day of gestation and the results were confirmed after the serial mating technique (as described in Materials and methods). The most prominent changes observed were (a) the disappearance of intense protein bands of approximately 78,000, 73,000, 71,000, 49,000, 43,000, 34,000, 28,000 Da from the supernatant fraction and 1000,000, 88,000, 26,000 Da from the pellet fraction and (b) the reduction of intensity of protein bands of approximately 75,000, 60,000, 51,000, 44,000, 38,000, and 32,000 Da from the pellet fraction and of approximately 265,000 Da protein from supernatant fraction.

3.2 Interpretation

Cyclophosphamide (CP) is an alkylating agent, and DNA synthesis is sensitive to the action of alkylating agents [26]. It has further been reported that single high dose treatment (50–100 mg/kg) of CP leads to a decreased DNA and protein synthesis in spermatids [27], followed by a decrease in testicular weight [28]. The present study showed that the number of proteins inhibited increased with the increasing dose. Only a single protein of 44 KD was apparently observed to be inhibited in the supernatant fraction of 2/12 experimental group, whereas 10 and 20 mg/kg doses of CP were sufficient to inhibit a variety of proteins. This difference in sensitivity and different results probably could be because of the difference in the nature of experimental material used in the two studies. It was observed that in 10/12 and 20/12 experimental group's proteins of 64, 58 and 44 KD were inhibited. Further, in the 20/12 experimental group protein (s) of approximately less than 20 KD molecular weight was also inhibited in the supernatant fraction. The results suggested that the low dose of CP (2 mg/kg) treatment did not affect the protein profile changes when compared to 10 and 20 mg/kg doses of CP exposure to developing testes, either due to very low amount of CP, or due to the recovery of the content during the period of embryonic life as the gestational age progressed. There was an apparent qualitative change in major protein profile observed in the testicular tissue from 1-day-old pups exposed antenatally to CP at 12th day of gestation. It has been observed earlier that CP affects mesenchymal tissues [29] and testes being mesenchymal in origin, it is quite possible that on 12th day of gestation when differentiation of male sex takes place [3, 30], this alkylating agent apparently suppressed over all protein synthesis and inhibited specific polypeptides of the testicular cells during development of gonads.

The protein profile of the pellet fraction did not show similar effects with respect to increasing dose as observed in the supernatant fraction. The apparent increase in the intensity of the protein bands in 20/12 experimental group could probably be because of various reasons such as the cells failed to replicate DNA after alkylating damage, which may have led to mitotic delay and cell enlargement associated with abnormal RNA and protein synthesis. It is also possible that cross linking of DNA, inhibition of protein synthesis and arrest of the cell-cycle in G_2 stage resulted in cell death. Although the exact mechanism of cell death has not been established, it could be possible that after the cell death the accumulation of protein (s) takes place due to apoptosis too.

The protein profile in testicular tissue from 1-day-old pups exposed antenatally to CP on 15th day of gestation caused less changes as compared to 12th day of gestation, probably because at this stage (15th day of gestation) testis is completely differentiated and a well-defined tunica albuginea appears (Fig. 3.2) [31].

The low dose of CP (2 mg/kg) did not affect the protein profile changes in the supernatant as well as the pellet fractions. The protein profile changes in pellet fraction showed the disappearance of 24 and 21 KD, decrease in the intensity of 58 and 44 KD in 10/15 experimental group. The protein band of 44 KD disappeared from

Fig. 3.2 T.S. of rat fetus at day 15 after H& E staining. Male sex—differentiation, testis being develop in close proximity to mesonephros *(→)* and double-layer tunic albuginea *(→)* covering formation of seminiferous tubules (→) inside the testis

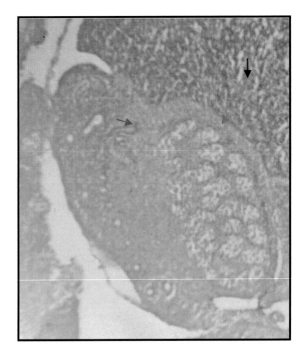

the supernatant fraction in 10/12, 20/12 and 10/15 experimental groups probably due to the fact that CP completely inhibited the protein and the recovery within 10-days or 7-days of interval after CP treatment did not take place, whereas the increased intensity of protein bands in 20/15 experimental group probably could be due to either the selective cellular death or the possible recovery within 7 days (i.e., the period between treatment of CP and collection of the tissue). It could be possible that due cell death degradation of protein takes place, resulting in an accumulation of abnormal amount of protein. It has been reported earlier that CP quantitatively affects the synthesis of DNA and protein in developing embryo [32].

The antenatal exposure of male pups to CP on 18th day of gestation, the time by which the testicular development is already over, did not show much inhibition of polypeptides but did not affect the intensity of protein bands, reflecting the possible alteration in the synthesis of polypeptides at relatively higher doses of CP. The lowest dose (2 mg/kg) of CP seemed to have effect on protein profile, whereas 10 and 20 mg/kg doses of CP also showed less changes in protein profile as compared to 10/12, 20/12 and 10/15 and 20/18 experimental groups, probably due to—(1) the low dose of CP to be unable to inhibit proteins, (2) the recovery of the inhibited protein during the course of development or (3) the requirement for more time for CP show its activity for cellular toxicity. It was observed earlier that inhibition of protein synthesis was not observed until 72 h after CP treatment [32], and it could be possible that CP required more time to metabolize or transfer the metabolite (s) to the target tissue for selective cell killing activity.

Present study includes that these proteins of molecular weight 86, 64 58, 44 KD and less than 20 KD from the supernatant fraction, and 74.5, 64, 58, 44, 24 and 21 KD from the pellet fraction of the testicular from 1-day-old pups exposed antenatally to CP in 2/12, 10/12, 20/12, 10/15, 20/15 and 10/18, 20/18 experimental groups were affected (i.e., disappeared or decreased or increased in intensity) after exposure with CP.

These proteins fell in the molecular weight range of 16.5–89 KD which appear at the time of spermatogenesis (Stern et al. [33]. It could be possible that these protein appear well before the initiation of spermatogenesis and play some role in spermatogenesis. It was further observed that the protein 24 KD which was observed in new born rat testes [3] was inhibited in 10/15 and 20/15 experimental group, whereas the protein of similar molecular weight decreased in intensity in 10/12 and 10/18 experimental group, probably suggesting that 24 KD protein observed in the embryonic tissue at 12th and 13th day of gestation and in the newborn is different. It could also be explained if the gene for this protein was active at two different times during embryonic development.

The embryonic material on 12th and 15th day of gestation, the period used for insulting the developing embryo in the already described results, apart from the days in-between i.e. 13th and 14th, were also used for protein profile studies (region of the embryos studied and the purpose is described in Materials and methods) to know if the proteins, inhibited in the testicular tissue from 1-day-old pups exposed antenatally to CP at 12th and 15th day of gestations appeared normally during these gestational periods.

The present study indicated that molecular weight ranges of the proteins from 146 KD to less than 20 KD appeared in the supernatant fraction, and 94 KD to less than 20 KD in the pellet fraction of normal male and female embryonic tissues collected at 12th to 15th day of gestation. The proteins of 58 KD which appeared in normal developing male embryo of day 12 and 13, disappeared in both the supernatant and pellet fraction of 20/12 experimental group. However, a protein of similar molecular weight showed decreased intensity in the supernatant fraction of 10/18 and the pellet fraction of 2/12 experimental group. A 44 KD molecular weight protein which appeared in normal developing embryos of both the sexes from 12th to 15th day of gestation disappeared in the supernatant fraction of 2/12, 10/12, 20/12 and showed a decrease in the intensity in 10/18 experimental group. The pellet fractions, however, showed the similar molecular weight protein with decreased intensity in 10/12, 10/18 and increased intensity in 20/12 experimental group.

Interestingly, it was further observed that the protein of less than 20 KD which appeared in normal developing male embryos of 12th and 13th days and the embryo of both the sexes from 14th to 15th days of gestation, disappeared in the supernatant fraction 20/12 and showed an increased intensity in 10/12 and 20/18 experimental groups. The above observations suggested that these proteins probably appeared at or before the time of the differentiation of gonad and any insult to the embryo resulting in inhibition of protein synthesis could lead to the infertility in future. It is also likely that the proteins with similar molecular weights produced in the embryo and in the adult are different in nature and physiological function. However, these conclusions are based on experimental design which may require for reconfirmation.

The qualitative changes in the protein profile were also observed in the testicular tissue obtained from adult infertile rats that were exposed antenatally to a continuous dose of CP of 5 mg/kg at 12th to 15th day of gestation. The results showed that the proteins of approximately 49 and 34 KD molecular weight disappeared from the supernatant fraction, whereas a protein of 44 KD showed decreased intensity in the pellet fraction. A similar molecular weight protein was observed to be affected in the experimental groups of 2/12, 10/12 and 20/12 (already described in the results). Again, the same protein (44 KD) was observed to appear in early embryonic development, and it is assumed that this protein might be playing an important role in fertility, because the similar molecular weight proteins (34, 49 and 44 KD) were found during the period of spermatogenesis (Stern et al. [33]. The results indicate that protein of 34 and 28 KD molecular weight disappeared from the supernatant fraction. The proteins of the same molecular weight have been reported by Shanowittz et al. [16] in seminiferous tubular fluid (SNF) and testicular intertubular fluid (TIF).

In our experimental study, it was also observed that high molecular weight proteins (110, 88 KD) disappeared from pellet fraction Carson [34] has reported that high molecular weight proteins are retinol binding which play an important role in spermatogenesis [19].

In conclusion, it is apparent that CP inhibits the synthesis of macromolecules (proteins) which have an important function in gene expression [35]. Also, it could affect the fertility of F_1 generation, although the exact mechanism of infertility is not clear, but it indicates that continuous insulting with CP at the time of gonadal differentiation causes abnormal development which leads to infertility in males.

References

1. J.W. Davidson, E.A. Clarke, Influence of modern radiological techniques on clinical staging of malignant lymphomas. Can. Med. Assoc. J. **99**(24), 1196–1204 (1968)
2. J.R. Davis, G.A. Langford, Comparative responses of the isolated testicular capsule and parenchyma to autonomic drugs. J. Reprod. Fertil. **26**(2), 241–245 (1971)
3. U. Muller, H. Schindler, W. Schempp, K. Schott, E. Neuhoff, Gene expression during gonadal differentiation in the rat. A two dimensional gel electrophoresis investigation. Dev. Gen. **5**, 27–42 (1984)
4. B.M. Sanborn, A. Steinberger, M.L. Meistrich, E. Steinberger, Androgen binding sites in testis cell fractions as measured by a nuclear exchange assay. J. Steroid Biochem. **6**(11–12), 1459–1465 (1975)
5. W.S. Kistler, M.E. Geroch, H.G. Williams-Ashman, Specific basic proteins from mammalian testes. Isolation and properties of small basic proteins from rat testes and epididymal spermatozoa. J. Biol. Chem. **248**(13):4532–4543 (1973)
6. N.C. Mills, A.R. Means, Sorbitol dehydrogenase of rat testis: changes of activity during development, after hypophysectomy and following gonadotrophic hormone administration. Endocrinology **91**(1), 147–156 (1972)
7. V.L. Go, R.G. Vernon, I.B. Fritz, Studies on spermatogenesis in rats. 3. Effects of hormonal treatment on differentiation kinetics of the spermatogenic cycle in regressed hypophysectomized rats. Can. J. Biochem. **49**(7), 768–75 (1971)
8. G.D. Hodgen, R.J. Sherins, Enzymes as markers of testicular growth and development in the rat. Endocrinology **93**(4), 985–989 (1973)
9. M. Hintz, E. Goldberg, Immunohistochemical localization of LDH-x during spermatogenesis in mouse testes. Dev. Biol. **57**(2), 375–384 (1977)
10. M.L. Meistrich, P.K. Trostle, M. Frapart, R.P. Erickson, Biosynthesis and localization of lactate dehydrogenase X in pachytene spermatocytes and spermatids of mouse testes. Dev. Biol. **60**(2), 428–441 (1977)
11. J.M. Kramer, R.P. Erickson, Analysis of stage-specific protein synthesis during spermatogenesis of the mouse by two-dimensional gel electrophoresis. J. Reprod. Fertil. **64**(1), 139–144 (1982)
12. M.K. Skinner, M.D. Griswold, Sertoli cells synthesize and secrete transferrin-like protein. J. Biol. Chem. **255**(20), 9523–9525 (1980)
13. W.W. Wright, N.A. Musto, J.P. Mather, C.W. Bardin, Sertoli cells secrete both testis-specific and serum proteins. Proc. Natl. Acad. Sci. U.S.A. **78**(12), 7565–7569 (1981)
14. R.M. DePhilip, A.L. Kierszenbaum, Hormonal regulation of protein synthesis, secretion, and phosphorylation in cultured rat Sertoli cells. Proc. Natl. Acad. Sci. U.S.A. **79**(21), 6551–6555 (1982)
15. C. Kissinger, M.K. Skinner, M.D. Griswold, Analysis of Sertoli cell-secreted proteins by two-dimensional gel electrophoresis. Biol. Reprod. **27**(1), 233–240 (1982)
16. R.B. Shabanowitz, R.M. DePhilip, J.A. Crowell, L.L. Tres, A.L. Kierszenbaum, Temporal appearance and cyclic behavior of Sertoli cell-specific secretory proteins during the development of the rat seminiferous tubule. Biol. Reprod. **35**(3), 745–760 (1986)
17. N.A. Musto, G.L. Gunsalus, C.W. Bardin, Purification and characterization of androgen binding protein from the rat epididymis. Biochemistry **19**(13), 2853–2860 (1980)
18. S.B. Wolbach, P.R. Howe, Nutrition classics. J. Exp. Med. **42**, 753–777 (1925); Tissue changes following deprivation of fat-soluble A vitamin. Nutr. Rev. **36**(1), 16–19 (1978)

19. H.F. Huang, W.C. Hembree, Spermatogenic response to vitamin A in vitamin A deficient rats. Biol. Reprod. **21**(4), 891–904 (1979)

20. D.S. Goodman, Overview of current knowledge of metabolism of vitamin A and carotenoids. J. Natl. Cancer Inst. **73**(6), 1375–1379 (1984)

21. D.B. McClure, S. Ohasa, G.H. Sato, Factors in the rat submaxillary gland that stimulate growth of cultured glioma cells: identification and partial characterization. J. Cell. Physiol. **107**(2), 195–207 (1981)

22. M. Kato, W.K. Sung, K. Kato, D.S. Goodman, Immunohistochemical studies on the localization of cellular retinol-binding protein in rat testis and epididymis. Biol. Reprod. **32**(1), 173–189 (1985)

23. S.B. Porter, D.E. Ong, F. Chytil, M.C. Orgebin-Crist, Localization of cellular retinol-binding protein and cellular retinoic acid-binding protein in the rat testis and epididymis. J. Androl. **6**(3), 197–212 (1985)

24. A.K. Saxena, R. Bamezai, Visualization and comparison of protein bands in the same SDS-PAG with simultaneous use of three different stains. Indian J. Exp. Biol. **26**(11), 866–868 (1988)

25. K. Weber, M. Osborn, The reliability of molecular weight determinations by dodecyl sulfate-polyacrylamide gel electrophoresis. J. Biol. Chem. **244**(16), 4406–4412 (1969)

26. J.J. Roberts, A.R. Crathorn, T.P. Brent, Repair of alkylated DNA in mammalian cells. Nature **218**(5145), 970–972 (1968)

27. I.P. Lee, R.L. Dixon, Antineoplastic drug effects on spermatogenesis studies by velocity sedimentation cell separation. Toxicol. Appl. Pharmacol. **23**, 20–41 (1972)

28. F. Pacchierotti, D. Bellincampi, D. Civitareale, Cytogenetic observations, in mouse secondary spermatocytes, on numerical and structural chromosome aberrations induced by cyclophosphamide in various stages of spermatogenesis. Mutat. Res. **119**(2), 177–183 (1983)

29. J.N. Singh, Q. Jehan, B.S. Setty, A.B. Kar, Effect of some steroids on spermatogenesis & fertility of rats. Indian J. Exp. Biol. **9**(2), 132–137 (1971)

30. A. Jost, S. Magre, R. Agelopoulou, Early stages of testicular differentiation in the rat. Hum. Genet. **58**(1), 59–63 (1981)

31. U. Mittwoch, J.D. Delhanty, F. Beck, Growth of differentiating testes and ovaries. Nature **224**(5226), 1323–1325 (1969)

32. R.D. Short, K.S. Rao, J.E. Gibson, The in vivo biosynthesis of DNA, RNA, and proteins by mouse embryos after a teratogenic dose of cyclophosphamide. Teratology. **6**(2), 129–137 (1972)

33. L. Stern, B. Gold, N.B. Hecht, Gene expression during mammalian spermatogenesis. 1: Evidence for stage-specific synthesis of polypeptide in vivo. Biol. Reprod. **28**, 483–496 (1983)

34. R.F. Carson, T.E. Batchman, Polarization effects in silicon-clad optical waveguides. Appl. Opt. **23**(17), 2985 (1984)

35. W.D. Davidson, C.A. Lemmi, J.C. Thompson, A study of the role of protein synthesis in the stimulation of acid secretion by gastrin and a gastrin-like pentapeptide. Endocrinology **82**(2), 416–419 (1968)

Chapter 4
Effect of Cyclophosphamide on Testis—A Histological Study

4.1 Introduction

In recent years, an increasing number of reports have appeared on the testicular damage following antimitotic drug therapy for the treatment of malignant conditions. In view of the marked cytotoxicity of the anticancer drugs, particularly alkylating agents on human and animal fertility [1, 2], hence, the study becomes relevant on testicular damage has been ascribed to several drugs such as cyclophosphamide [3–5]. Various reports on the subject show that high dose of cyclophosphamide (CP) during specific, days critical period of pregnancy can be teratogenic in experimental animals [6–8]. The effects of these cytotoxic drugs on the testis have been studied by testicular biopsy, semen analysis and endocrine assessment of the hypothalamo-hypophyseal testicular axis. However, the literature lacks adequate mention regarding the effect on the germinal and non-germinal elements (i.e., gonocytes and sertoli cells) in the testis of 1-day-old rats exposed in utero (antenatally) to different doses of CP.

As CP is an alkylating agent which causes maximum damage to rapidly dividing cells, the drug was expected to affect the seminiferous epithelium to reduce the number of spermatozoa produced [9]. The seminiferous epithelium in the fetal testis shows two distinct cell types which include gonocytes (primordial germ cells) and the supporting cells, precursors of the sertoli cells. During fetal life, testis shows active mitotic activity and testicular cords undergo conspicuous morphological and biochemical changes to form the seminiferous tubules of the adult testis. Merchant [10] reported that the number of sertoli cells and germ cells increases in tubules of rat, and these two cell lines, germ cells and sertoli cells, continue to be present in the seminiferous tubules of developing and maturing testis [10–16].

The CP-induced testicular damage has been observed dose-dependent manner, and high doses of CO destroyed all stages of spermatogenic elements in the rat [3, 5, 17]. Germinal cell aplasia ('Sertoli cell-only' syndrome) was induced by a variety of methods: administration of busulfan, hydroxyurea and vitamin A-deficient diet during embryonic life [18–21]. In fetal testicular cord, the germ cells are called gonocytes, randomly distributed as single large cells among supporting cells (sertoli

© The Author(s), under exclusive license to Springer Nature Singapore Pte Ltd. 2020
A. K. Saxena and A. Kumar, *Fish Analysis for Drug and Chemicals Mediated Cellular Toxicity*, SpringerBriefs in Applied Sciences and Technology,
https://doi.org/10.1007/978-981-15-4700-3_4

cells). Their cytoplasm appears clear than the surrounding cells [10, 22, 23]. The primary movement of the gonocytes seems toward the periphery of the seminiferous tubule [15] leading to eventual disappearance of centrally located germ cell. It was observed that the number of gonocytes in mouse testes increased significantly after ethinyl estradiol treatment and decreased with irradiation [24–26].

CP has also been associated with qualitative changes of sertoli cells. Previous studies have shown that the sertoli cells partially control the development of seminiferous epithelium and spermatogenetic processes of mammals [27]. The origin of sertoli cell is not known, and certain mesenchyme, surface epithelium and mesonephric tubules have all been suggested as possible sources. Although none of these can be excluded at the present time, morphological evidence suggests epithelium origin. The columnar arrangement of the sertoli cells and their initial ultrastructural resemblance to surface epithelium cells [28] also point epithelial rather than mesenchymal origin. Whereas, the positive correlation between the total number of sertoli cells, the total length of seminiferous tubules per testes and the total areas of seminiferous epithelium have been observed area increased after ethinyl estradiol treatment in developing testes [26], whereas the number of sertoli cells decreased in hypophysectomized immature rats [16, 27]. However, the qualitative relationship is developed between the sertoli cells per testes and the number of renewing spermatogonia per testis in adult rat. This provides an evidence of qualitative regulation of adult spermatogenesis by the population of sertoli cells during the inpubertal period of testicular development. It has been reported that sertoli cells play an important role in the early maturation of the seminiferous epithelium, and the site of conversion of progesterone to testosterone reported that number of sertoli cells per cross section and the ratio of the number of sertoli cells per gonocytes were significantly decreased in mouse testes after ethinyl estradiol treatment [26, 29].

It is reported that alkylating agents also influence the function of leydig cells. In the developing and maturing testes, the study of differentiation of the leydig cells is complicated due to the fact that not all the cells differentiate at the same time [28], resulting in heterogeneous population of cells. The leydig cells are the site for the synthesis of male hormones in the fetal testis and are also believed to influence the multiplication and differentiation of germ cells. The testicular development from 14.5 to 20.5 days is characterized by an increased capacity to synthesize testosterone and respond to L/H hormone. The decrease in testosterone production in the neonatal testis is because of reduction in the number of leydig cells with increasing age [16]. However, certain drugs, which are cytostatic in nature, show direct or indirect action on the seminiferous epithelium during early stages of development of seminiferous tubules and destroy germ cells [30, 31].

Drugs affecting hormonal pathway attack sertoli cells and at later stages of the seminal epithelium. CP has been associated with various disorders of spermatogenesis by changing the concentration of inhibin (FSH-suppressing hormone) secreted by sertoli cells, and with the increase of FSH and LH levels, resulting in tubular damage [2], however reported that in rats, lethal or near lethal quantities of nitrogen mustard (CP metabolite) were necessary to influence spermatogenesis and fertility, whereas previous studies indicated that the prolonged exposure for more than four weeks could be lethal (Wheeler et al. 1962) and the fertility was affected due to dominant lethal mutagenic effect [17, 32, 33]. It was observed that the chronic administration of low dose of cyclophosphamide had minimal effects on the male reproductive system and fertility, but resulted in malformation and growth retardation in the surviving fetus with a high frequency of fetal death [9].

However, the relevance of the effects of such chemicals on the male reproductive system of the offspring is poorly understood. In preview of these facts that the experimental plan was designed with the aim to investigate the effect of CP on developing testes. Such study would be helpful in knowing the mechanism of action of the drug on cytotoxicity during prenatal life and possibly correlate with reproductive performance of F_1 generation male rats.

4.2 Experimental Procedure

4.2.1 Materials

The testis of 1 day of old pups was collected from nine parallel experimental sets in which mothers were exposed (i.p.) with cyclophosphamide (i.e., single dose of 2 or 10 or 20 mg/kg) either on 12th or 15th or 18th day of gestation period. The testes were fixed in Bouin's solutions for 24 h. After fixation, the testes were dehydrated with ascending grade of alcohol and cleared in xylene. The testes were impregnated with paraffin wax (melting point, 56°–58°). The transverse sections of 7 μm were used after staining with hematoxylin and eosin.

Hematoxylin (Ehrlich's)

Absolute alcohol	33 ml
Hematoxylin	0.67 g
Glacial acetic acid	3.3 ml
Glycerol	33 ml
Distilled water	33 ml
Potassium alum	3 g

These were kept for ripening, for two months until a deep red color appears.
Eosin

Eosin	1.0 g
Ethyl alcohol	70–1000 ml
Glacial acetic acid	5 ml

Dilute with equal volume of 70% alcohol before use, and then add 2–3 drops of acetic acid.

Paraffin was removed after dipping the slide in xylol. Sections were brought to water through descending grades of alcohol. Hydrated sections were then stained with Ehrlich's acid hematoxylin for 10 min. Sections were blued in top water. These were count stained with eosin for 1 min—then, these were dehydrated through ascending grades of alcohol, cleaned in xylol and mounted in D.P.S.

Nucleus—bluish black

Cytoplasm—pink.

Counting of Sertoli cells and Gonocytes

The contents of the seminiferous tubule cross sections of both the experimental and control testes were germinal elements, the gonocyte and non-germinal elements, the sertoli cells. The gonocytes were distinguishable from the sertoli cells by their large size and by a round nucleus containing a large nucleolus. According to Gondo (1974) when no germ cells in tubule cross section show mitotic division, they are classified as prespermatogonia or spermatogonia, and were also counted as gonocytes. The sertoli cells could be easily distinguished from the germ cells, and they showed a columnar shape with an elongated nucleus and excentric, small and compact nuclei. At least five sections from each testis both experimental and controls, separated by 70 mm, were selected, and in each section ten seminiferous tubules were selected for counting the number of sertoli cells and gonocytes.

Measurement of This Area Occupied by the Seminiferous Tubules and Interstitial Space

The seminiferous tubular area and interstitial space were measured by using micrometer disk having 100 squares of size 5 mm^2, which was coincided with mm scale. Micrometers 2 mm long geteilt in 200 tilies, which comes to finally 0.64 mm^2. The total 15 sections were selected from experimental as well as control testes for the measurement of the area. The total number of empty squares was counted within the region of square disk under 6 × 40 magnifications. The empty squares denote the intertubular space, and rest of the square occupied by the tubular region indicate the seminiferous tubular area.

Measurement of the Size of the Leydig Cells

The diameter of the leydig cells was measured by using micrometer scale in eye piece. A total of 10 sections in each testis were chosen for the study, and in each section five cells were observed for the measurement of the diameter of the leydig cells. The size of the cells was measured by using grid of 0.015 mm scale at 6 × 40 magnifications.

Statistical Analysis

The histological features were observed in nine parallel sets of experiments, and it was proposed to find out the significant difference between three different doses (2 or 10 or 20 mg/kg) of CP at 12th or 15th or 18th day of gestation. In the testicular tissue to compare the differences between days and dose, the two-way analysis of variance (ANOVA) table was computed for the total number of sertoli cells/tubule. Cross section, Gonocytes/tubule cross section. Size of the Leydig cells and ratio of sertoli cell/gonocyte in tubule cross section, tubular region and interstitial space.

The f-values obtained and the significant point for the calculated f-values were obtained by Fisher and Yates table against degree of freedom 2 and 4. When both the calculated f-values were more than the tabulated f-values, it was indicated that the difference between days and dose was significant. In case the f-value was significant, the interaction was considered first, suggesting the effect of days, dose combination varies from day (or dose to dose). Therefore, using the critical difference comparison for significance level were made in the observed mean values of sertoli cell, gonocyte, ledig cell, tubular region and interstitial space. The data of the ratio of number of sertoli cells per gonocytes; number of sertoli cells; number of gonocytes per tubule cross section; seminiferous tubular area and intertubular space; and diameter of the leydig cells were analyzed by Student test.

Photomicrography
The photomicrographs for histological studies were taken with the help of Leitz Orthoplan (microscope) under maximum × 1024 magnifications. Negative film NP-40 documentation films were used and developed in fine grain developer, and the prints were made on Agfa extra white glossy hard printing paper.

In subhuman species, a particular cellular association occupies a relatively long segment along the tubule. Therefore, in a typical cross section only a single association is observed but in man cellular association occurs in irregularly shaped areas along the tubule, and therefore, a cross section typically shows two or more cellular associations.

4.3 Findings

The cellular features based on histology in both tubular and non-tubular regions (interstitial space) of the testes studied from 1-day old pups exposed antenatally to 2 or 10 or 20 mg/kg dose of CP at 12 or 15 or 10 days of gestation showed both qualitative and quantitative changes. These changes were:

Sertoli Cells
In the control testes, the average number of sertoli cells per tubule cross section was found to be 16.00 ± 1.88 whereas in experimental tests number decreased significantly ($p < 0.001$) in 20/12 (13.36 ± 2.74), 20/15 (14.08 ± 2.15) and 20/18 (11.52 ± 1.63) experimental groups as shown in Fig. 4.1a, b. The number of sertoli cells also decreased significantly ($p < 0.001$) in 2/18 (13.72 ± 2.01) and 10/18 (13.16 ± 3.48) experimental groups (Table 4.1), when compared with controls. The f-values observed for days (31.29) and dose (89.49) were significantly higher than table f-value at 1% level (Table 4.2).

Gonocytes
The average number of gonocytes per tubule cross section was 1.98 ± 0.95 in the control testes. The number of the gonocytes decreased significantly ($p < 0.001$) with

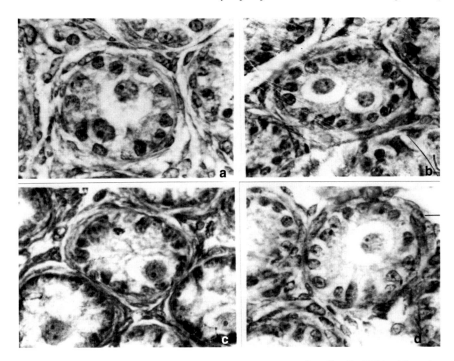

Fig. 4.1 **a** Seminiferous tubule in a 1-day old control rat pups. Sertoli cells (SCs) with oval or elongated nuclei are seen arranged in a uniform layer on the basement membrane. The gonocytes (GN) showing large oval nuclei and clear cytoplasm abundant in quantity. The nucleus shows many nucleoli. H & E × 1024. **b** Seminiferous tubules in a 1-day old control rat pups exposed to 2 mg/kg dose of CP on day 12 of gestation. The Sertoli cells (SCs) with elongated or oval nuclei as seen in a single uniform layer on the basement membrane. The gonocytes (GN) placed in the center of tubule are like those seen in control. **c** Seminiferous tubule in 1-day old rat pups exposed to 10 mg/kg of CP on day 12 of gestation. The sertoli cells (SCs) are less in number, and smaller size of nuclei from round to flattened basement membrane (BM) was interrupted (P_t). Gonocytes (GN) are seen migratory toward the periphery. **d** Seminiferous tubule in a 1-day old rat pups exposed to 20 mg/kg of CP on day 12 of gestation. The Sertoli cells (SCs), which are less in number per tubule, show nuclei with margin not well defined and merging with cytoplasm. Shape of nuclei is irregular and so their arrangement on basement membrane (BM). The gonocytes (GN) are seen near the basement membrane (BM) which is interrupted (P_t)

respect to control in 20/12 (1.38 ± 0.80), 20/15 (1.28 ± 0.96) and 20/18 (1.46 ± 0.81) experimental groups (Table 4.1). The observed f-values for days (10.14), doses (123.57) and interaction (50.42) were significantly higher than table f-values at level (Table 4.2). The mean difference between days and dose was observed and compared with critical difference. This showed that 15th day of the gestation was susceptible period for the development of gonocytes in 3 or 10 or 20 mg/kg doses of CP. The 20/18 experimental group also showed similar susceptibility affecting the number of gonocytes as shown in Fig. 4.1c, d.

Table 4.1 Effect of cyclophosphamide on tubular contents (number of Sertoli cells and gonocytes)/tubule cross section

Gestation period (days)	Dose of CP (mg/kg)	No. of sertoli cell/tubule cross section mean ± S.D	Significance level	Number of sertoli cell per tubule cross section mean ± S.D	Significance level
Control	–	16.00 ± 1.88		1.98 ± 0.95	
12	2	15.26 ± 2.49	$P > 0.05$	2.10 ± 1.11	$P > 0.05$
	10	14.82 ± 2.60	$P < 0.01$	2.06 ± 1.01	$P > 0.05$
	20	13.36 ± 2.74	$P < 0.001*$	1.38 ± 0.80	$P < 0.001*$
15	2	16.84 ± 2.16	$P < 0.05$	1.80 ± 0.92	$P < 0.05$
	10	16.06 ± 3.37	$P > 0.05$	1.58 ± 0.54	$P < 0.05$
	20	14.08 ± 2.15	$P < 0.001*$	1.28 ± 0.96	$P < 0.001*$
18	2	13.72 ± 2.01	$P < 0.001*$	1.66 ± 0.78	$P > 0.05$
	10	63.16 ± 3.48	$P < 0.001*$	1.72 ± 0.88	$P > 0.05$
	20	11.58 ± 1.63	$P < 0.001*$	1.46 ± 0.81	$P < 0.001*$

CP cyclophosphamide. *Statistically analysis showing highly ($p < 0.001$) significant difference with respect to controls

Table 4.2 Effect of different doses of cyclophosphamide on different days, on sertoli cells and gonocytes

Number of sertoli cells					Number of gonocytes		
Source	d.f.	Sum of the squares	Mean sum of the squares	f-value	Sum of the squares	Mean sum of the squares	f-value
Days	2	619.72	309.86	131.29**	15.42	7.71	110.14**
Doses	2	422.41	211.20	89.49*	17.30	8.65	123.57**
Interaction	4	9.67	2.41	1.02	14.14	3.53	50.42**
Error	441	1042.13	2.36		32.72	0.07	
Total	449	20.93.93			79.58		

		*		**	
		5%		1%	
2 d. f. & 441		3.02		4.660	
2 d. f. & 441		2.40		3.3	

ANOVA showing (f-values) a significant difference

Ratio of Sertoli Cells Per Gonocytes

The number of sertoli sells per gonocytes observed to be 9.96 ± 4.72 in the control testes significantly decreased ($p < 0.05$) in 20/12 (7.87 ± 5.07) and 20/18 (8.11 ± 4.33) experimental groups (Table 4.3). The f-values observed for days (182.30),

Table 4.3 Effect of cyclophosphamide on tubular contents (the number of sertoli cells and gonocytes) per tubule cross section

Number of Sertoli cells					Number of gonocytes		
Source	d.f.	Sum of the squares	Mean sum of the squares	f-value	Sum of the squares	Mean sum of the squares	f-value
Days	2	353.69	176.84	182.30**	1.33	0.66	110.14**
Doses	2	78.43	39.21	40.42**	2.33	1.16	123.57**
Interaction	4	146.76	39.69	37.82	0.27	0.06	50.42**
Error	441	432.12	0.97		3.66	0.008	
Total	449	1011.03			7.59		

doses (40.42) and interaction (37.82) were significantly higher than table f-values at 1% level. The mean differences showed 18th day of gestation period to be most sensitive period to all the doses (i.e., 2 or 10 or 20 mg/kg) of CP in developing testes.

In the non-tubular region, these changes were:

Leydig Cells

The single dose of CP also altered the size (diameter, mm^2) of the leydig cells in 1-day old testes (Table 4.4). The mean diameter of the Leydig cells in 10/12 (17.51 ± 2.68) and 20/12 (17.51 ± 2.69) experimental groups significantly decreased (p < 0.001) as compared to control (20.80 ± 3.07). In experiments of 10/15 (17.98 ± 2.24) and 20/15 (16.85 ± 1.98), also the size of Leydig cells with respect to controls significantly decreased. The mean diameter of the leydig cells in comparison with control significantly decreased (p < 0.001) in 2/18 (17.67 ± 2.79), 10/18 (16.58 ± 1.86) and 20/18 (15.50 ± 2.15) experimental groups (Table 4.5).

The f-values observed for days (32.50), dose (145.00) and interaction (7.50) were found significantly higher than table value at 1% level (Table 4.4), and the significant mean difference was also observed between 20/15 and 18/20 experimental groups when compared with critical differences.

Interstitial Space and Seminiferous Tubular Area

An increase of the interstitial space and a subsequent decrease in the seminiferous tubular area were observed in all the nine sets of experimental groups, i.e., 2/12, (58.30 ± 0.98), 10/12 (56.33 ± 3.15), 20/12 (37.76 ± 2.88); 2/15 (52.88 ± 1.54), 10/15 (54.36 ± 2.05), 20/15 (56.89 ± 0.56); and 2/18 (53.76 ± 4.14), 10/18 (54.33 ± 1.72), 20/18 (58.56 ± 1.10), from 1-day old pups exposed antenatally to CP (Table 4.6). The mean interstitial space in the control testes was 1.74 ± 0.67. In the experimental groups, the interstitial space of the testes gradually increased and the seminiferous tubular area decreased in 2/15 (11.11 ± 1.50), 10/15 (9.63 ± 2.05), 20/15 (7.10 ± 0.56), and 2/18 (10.24 ± 4.14), 10/18 (9.45 ± 1.64), 20/18 (5.42 ± 1.10) experimental groups. The maximum increase (26.19 ± 3.17) of interstitial space and the maximum decrease (37.76 ± 2.88) in seminiferous tubular area were observed in 20/12 experimental group. The density of the tissue in the interstitial

Table 4.4 Effect of different doses of cyclophosphamide on different days on the ratio of sertoli cells per gonocyte and size of leydig cells

Number of Sertoli cells					Number of gonocytes		
Source	d.f.	Sum of the squares	Mean sum of the squares	f-value	Sum of the squares	Mean sum of the squares	f-value
Days	2	353.69	176.84	182.30**	1.33	0.66	110.14**
Doses	2	78.43	39.21	40.42**	2.33	1.16	123.57**
Interaction	4	146.76	39.69	37.82	0.27	0.06	50.42**
Error	441	432.12	0.97		3.66	0.008	
Total	449	1011.03			7.59		

		*		**	
			5%		1%
2 d. f. & 126			3.02		4.66
4 d. f. & 126			2.40		3.37

ANOVA (f-value) showing a significant difference

Table 4.5 Effect of cyclophosphamide on the size of the Leydig cells

Gestation period (days)	Dose of CP (mg/kg)	No. of sertoli cell/tubule cross section mean ± S.D	Significance value
Control	–	20.30 ± 3.67	–
12	2	20.77 ± 4.04	$p > 0.05$
	10	17.51 ± 2.68	$P < 0.001**$
	20	17.51 ± 2.69	$P < 0.001**$
15	2	19.37 ± 3.98	$P > 0.05$
	10	17.98 ± 2.94	$P < 0.001**$
	20	16.85 ± 1.98	$P < 0.001**$
18	2	17.67 ± 2.79	$P < 0.001**$
	10	16.58 ± 1.86	$P < 0.001**$
	20	15.50 ± 2.15	$P < 0.001**$

**Statistical analysis showing highly ($p < 0.001$) significant difference with respect to controls

space was more compact in the control testes, whereas in the experimental groups the mesenchyme appeared less compact.

The f-values observed for days (35.11), dose (27.91) and interaction (110.52) were found significantly higher than table value at 1% level in case of intertubular space whereas, in case of tubular region, the f-values for days (35.04), dose (27.87) and interaction (110.01) also showed significant difference (Table 4.7).

Table 4.6 Reduction in the area occupied by seminiferous tubule with subsequent increase in the interstitial space following cyclophosphamide treatment

Gestation period (days)	Dose of CP (mg/kg)	Tubular area (mm^2) mean ± S.D	Significance as compared to control	Interstitial space (mm^2) mean ± S.D	Significance as compared to control
Control	–	62.25 ± 0.67		1.74 ± 0.67	
12	2	58.30 ± 0.98	$P < 0.001$	5.69 ± 0.99	$P < 0.001$**
	10	56.33 ± 3.15	$P < 0.001$	7.63 ± 2.05	$P < 0.001$**
	20	37.76 ± 1.54	$P < 0.001$	26.19 ± 3.17	$P < 0.001$**
15	2	52.88 ± 1.54	$P < 0.001$	11.11 ± 1.50	$P < 0.001$**
	10	54.36 ± 2.05	$P < 0.001$	9.63 ± 2.05	$P < 0.001$**
	20	56.89 ± 0.56	$P < 0.001$	10.24 ± 0.56	$P < 0.001$**
18	2	53.76 ± 4.14	$P < 0.001$	10.24 ± 4.14	$P < 0.001$**
	10	54.33 ± 1.72	$P < 0.001$	9.45 ± 1.64	$P < 0.001$**
	20	58.56 ± 1.10	$P < 0.001$	5.42 ± 1.10	$P < 0.001$**

**Statistical analysis showing highly ($p < 0.001$) significant difference with respect to controls

Table 4.7 Effect of different doses of cyclophosphamide on different days on intertubular space and tubular region

For intertubular space					For tubular region		
Source	d.f.	Sum of the squares	Mean sum of the squares	f-value	Sum of the squares	Mean sum of the squares	f-value
Days	2	1433.49	716.74	35.11**	1440.30	720.15	35.04**
Doses	2	1139.33	569.66	27.91**	1145.50	572.75	27.87**
Interaction	4	9023.26	2255.81	110.52**	9043.62	2260.90	110.01**
Error	126	2572.82	20.41		2589.80	20.55	
Total	134	14,168.90			14,219.22		
			*			**	
			5%			1%	
2 d. f. & 126			3.07			4.78	
4 d. f. & 126			2.44			2.47	

ANOVA (f-value) showing a significant difference

Basement Membrane

An apparent observation of the control testes showed a well-formed basement membrane, continuous in nature which surrounded the whole length and periphery of the seminiferous tubule (3.4, 3.5 and 3.6 A). whereas in experimental testes, the basement membrane was found to be discontinuous at several places and was not completely surrounding the seminiferous tubule, especially in 10/12, 20/12; 10/15, 20/15; and 10/18, 20/18 experimental groups.

4.4 Tunica Albugenia

There was no apparent difference in thickness of *Tunica albugenia* between 2/12, 10/12, 2/15, 10/15 and 2/18, 10/18 experimental groups and the controls. The thinning of the *T. albugenia* was observed in higher doses (i.e., 20/12, 20/15 and 20/18) of CP. However, the maximum thinning of the *T. albugenia* was observed in 20/12 experimental group when compared to the controls.

4.5 Interpretation

The present study includes one-day old rat testis, exposed antenatally to different doses either 2 or 10 or 20 mg/kg of cyclophosphamide (CP) on either 12th or 15th or 18th day of gestation to evaluate dose-related quantitative changes in tubular contents, ratio of sertoli cells per gonocytes, area in a cross section of tubule occupied by seminiferous tubules and the interstitial space in the cross section of the testis not occupied by tubules. In the qualitative change, the size of the leydig cells was found to be affected. Other changes observed were in the nature of the basement membrane of seminiferous tubules and the thickness of the *T. albugenia.*

The present study revealed that the number of sertoli cells decreased in the seminiferous tubule with increasing dose of CP. The maximum decrease in the number of sertoli cells in seminiferous tubule was observed with 20 mg/kg dose of CP at all the three gestations. It was also observed that even lower doses of CP (i.e., 2 and 10 mg/kg) caused significant decrease in the number of sertoli cells when fetuses were exposed antenatally on 18th day of gestation, whereas normally the number of sertoli cells increases throughout the fetal period due to continuous mitotic activity [16]. The decrease in the number of sertoli cells in the present study even at lower doses could be due to the antimitotic effect of CP, further indicating that day 18 of gestation was most sensitive period in the differentiation and development of sertoli cells. The time interval between the day of treatment and the day of collection of testicular tissue was very important in addition to the dose of CP. This was clear from negligible effect of smaller doses at early gestations (2 and 10 mg on days 12 and 15) which could probably be explained on the basis of enough available time interval for the recovery of sertoli cells from the assault by CP. It is likely that the specialized morphology of supporting cells could be observed at 16th day of gestation, and the seminiferous tubules are composed of precursors of sertoli cells and gonocytes which further differentiate to give rise to the fetal Sertoli cells from day 17 to day 19 of fetal life [10]. The experimental study indicated that 18th day of gestation period was the most sensitive period as all the doses of CP (i.e., 2 or 10 or 20 mg/kg) led to significant decrease in the number of sertoli sells in seminiferous tubules of the developing testes and thus could be considered as critical period in the differentiation of the sertoli cell in rat. The numbers of sertoli cells are also known to decrease in developing mouse testes after ethinyl estradiol treatment and hypophysectomized rat in prepubertal stage. The decrease in number of sertoli cells suggests it as a good parameter for estimating the effect of such drugs on rat testis [26].

A relationship was also observed between sertoli cells and germ cells during fetal development. The results of the present study indicated that ratio of sertoli cells per gonocyte in the seminiferous tubule decreased significantly ($p < 0.05$) on 18th day of gestation with 10 and 20 mg doses of CP again suggesting that 18th day of gestation period was the most sensitive period for disturbing the ratio of sertoli cells per gonocyte in seminiferous tubule and could be considered as a critical period for developing testes. The similar findings were also observed in developing mouse testes after ethinyl estradiol treatment [26]. A positive correlation between total number of sertoli cells per testis and the total length of seminiferous epithelium has been observed in rat. These quantitative relationships demonstrated between the number of sertoli cells per testis and the numbers of renewing spermatogonia per testes in adult rat have provided evidence of quantitative regulation of spermatogenesis by the population of sertoli cells which gets established during the prepubertal period of testicular development. Sertoli cells partially control the development of seminiferous epithelium and spermatogenesic processes in mammals. It is generally believed that sertoli cells have important roles in the early maturation of the seminiferous epithelium. The present study showed that high dose of CP caused discontinuity of seminiferous basement membrane in 20/12, 20/15 and 20/18 experimental groups, probably due to decrease in number of sertoli cells in the above experimental groups, as sertoli cells partially control the development of the seminiferous epithelium and a positive correlation between the total number of sertoli cells and the total area of the seminiferous epithelium has been observed in adult rat [16, 27].

In the fetal testicular cords, the germ cells which are generally called gonocytes are randomly distributed as single large cells among many sertoli cells precursors. Various studies have shown that the gonocytes in the rat have their regular round to oval shape and a large certainly placed spherical nucleus with evenly distributed chromatin and multiple nucleoli [10, 22]. Their cytoplasm appears clearer than that of supporting cells. The similar findings were also observed in our experimental studies. The previous studies of germ cells in the fetal and postnatal testes of rat have clearly demonstrated the continuity in the development and differentiation of fetal gonocyte to adult spermatocyte [11, 34].

During development of seminiferous tubules, the movement of gonocytes toward the basal lamina has been suggested by various workers [15]. Similar findings were also observed in our experiments which progressively increase toward peripherally located germ cells and the eventual disappearance of centrally located germ cells, and the primary movement of gonocytes appeared to be toward the periphery.

The mechanism by which the germ cells were shifted adjacent to the basal lamina is still not clear. The movement of germ cells toward the periphery is believed to have a protective function. However, some peripheral cells also undergo degeneration. In rat, the gonocyte multiplication stops at 17–18 days of gestation [13, 24] and is followed by a quiescent period lasting 1st week until the fourth postnatal key when type A spermatogonia first appears. Earlier reports showed that gonocytes of the developing fetus were sensitive of busulfan at 13th day of gestation and sensitivity of gonocytes decreased rapidly after birth [2]. The results of present study showed that highest dose of CP (20 mg/kg) caused significant decrease in the gonocytes in seminiferous tubule when fetuses were exposed antenatally at 12th or 15th or 18th day of gestation. The 15th day of gestation, however, was found to be most sensitive period in the differentiation of a gonocyte, as even the lowest dose of CP (10 mg/kg) caused decrease in the number of cells (gonocytes) significantly. Therefore, 15th day of gestation could be considered as a critical period in the development of gonocytes. It has been observed by earlier workers that gonocytes first appear on 15th day of gestation in the sex cord [10, 22]. However, it is quite possible that CP is especially more effective at that time when cell shows its first appearance, and CP checks the multiplication of rapidly dividing cells (gonocytes), resulting in the significant decrease of gonocytes' number in seminiferous tubule. Similar findings were also observed [26] in developing mouse testes after ethinyl estradiol treatment.

The movement of the germ cells to the periphery brings them nearer to the testosterone producing interstitial leydig cells, possibly an important prerequisite for further maturation. The maturation of gonocytes to spermatogonia might be under the control of testosterone [35, 36]. In the developing testis, the interstitial leydig cells resemble partly the undifferentiated fibroblast-like cells and partly the fully differentiated and mature interstitial leydig cells [10, 16, 28] (Merchant 1977). It has been now well established that the testicular interstitial leydig cells of the mature males develop through intermediate stage from 'fibroblast-like' interstitial cells of mesenchymal origin. The study of leydig cells in developing testis is complicated by the fact that not all the cells differentiate at the same time. This results in a heterogeneous population of leydig cells at different stages of their differentiation. The results of present study also revealed that the size of the leydig cells varied at different stages of testicular developing and the 18th day of gestation was most sensitive to CP in this respect. At this stage, even the minimum dose of CP (i.e., 2 mg/kg) altered the size of the leydig cells and the decrease was significant probably due to antimitotic nature of this drug. Therefore, 18th day of gestation period could be considered as a critical period also in the development of leydig cells. Earlier reports suggested that in fetal testis typical leydig cells were developed around day 17 of gestation and maximum development occurred of fetal day 19 [10, 37]. The above finding supports the results of prenatal exposure at 18th day of gestation period which was most sensitive period for the size of the leydig cells. Probably, this is the time for maximum growth attained by the leydig cells. In that study, it was reported that immediately after birth, leydig cells showed an abrupt decrease in number but after day 4 of postnatal life, there was a slow decline in their number [26]. The leydig cells are the site for the synthesis of male hormone in the fetal testis as revealed by various biochemical findings [38]. During the androgens synthesis in and released from embryonic and fetal leydig cells are also believed to influence the multiplication and differentiation of germ cells [16].

The developing of the seminiferous tubule has been studied in detail in different mammalian species [35]. The findings of the present study indicated that the area occupied by seminiferous tubules in a cross section of testis decreased after increase of CP dose and maximum reduction was observed in 20/12 experimental group followed by subsequent increase of interstitial space in testicular tissue, whereas the minimum changes with respect to interstitial space were observed in 2/12 experimental group, probably due to either low amount of drug or low metabolic activity which could not affect the seminiferous tubular region and interstitial space as compared to the higher doses of CP. The seminiferous epithelium in the fetal testis shows two distinct cell types, the gonocytes and sertoli cells, which show active mitotic activity during fetal life. As seen in the present study, also number of sertoli cells and gonocytes decreased in seminiferous tubule after CP treatment and the maximum effect was observed in 20/12 experimental group. This decrease in number of dividing cells possibly led to reduction in diameter of the seminiferous tubule due to reduction of tubular contents, resulting in subsequent increase of interstitial space. It was reported that number of germ cells and sertoli cells increased per cross section, and from 17th to 19th day of fetal life, the tubule attained a diameter of 46–50 μm [16]; dichloroacetydaimines in adult rat testis has also been shown to cause decrease in the tubular diameter due to pyknosis, karyolysis or karyorrhexis resulting in increased tubular vacuolization [39]. The maximum thinning of the *T. albugenia* was observed in 20/12 experimental group as compared with control testis; however, the study of Zschauer and Hodel [31] reported that tunica becomes thicker after treatment with ethinyl estradiol in adult rat testis. The differences in the results could probably be due to the different natures of chemicals used. In the experimental testis, tunica was less compact, with large mesenchymal areas observed just deep to the tunica, suggesting defective and poor development of the *T. albugenia*, at 12th day of gestation.

The following conclusion could be drawn from the findings of the present study that the CP interferes with the proliferation of the cellular elements in the seminiferous tubules of rat testis and leads to the qualitative and qualitative reduction of such elements resulting in diminished tubular diameter most evident on day 12 of treatment.

The testicular changes in 1-day old pups exposed antenatally to CP are dose-dependent. Day 18 of fetal development can be considered as the critical period in the developmental process of the sertoli cells and leydig cells, whereas day 15 can be considered as the critical period in the developmental process of gonocytes.

Stunting of the sertoli cell and gonocytes has been a striking observation in present study, especially with respect to a higher dose o CP, probably because of the cytostatic nature of the drug. The present study indicates that the number of gonocytes decreases significantly after CP administration and size of the leydig cells also gets altered which probably could affect the synthesis of testosterone and further maturation of gonocytes, presumably leading to altered fertility in F_1-generation of males; therefore, it is quite possible that delayed fertility observed in 20/12 experimental group probably was due to the effect on leydig cells interfering with testosterone synthesis. The reduction of tubular contents, both germinal and non-germinal elements, in developing testes could also affect the fertility in males.

References

1. S.M. Shalet, Effects of cancer chemotherapy on gonadal function of patients. Cancer Treat. Rev. **7**(3), 141–152 (1980)
2. H. Jackson, The effects of alkylating agents on fertility. Br. Med. Bull. **20**, 107–114 (1964)
3. M.S. Qureshi, J.H. Pennington, H.J. Goldsmith, P.E. Cox, Cyclophosphamide therapy and sterility. Lancet **2**(7790), 1290–1291 (1972)
4. G.L. Warne, K.F. Fairley, J.B. Hobbs, F.I. Martin, Cyclophosphamide-induced ovarian failure. N. Engl. J. Med. **289**(22), 1159–1162 (1973)
5. M. Lendon, I.M. Hann, M.K. Palmer, S.M. Shalet, P.H. Jones, Testicular histology after combination chemotherapy in childhood for acute lymphoblastic leukaemia. Lancet **2**(8087), 439–441 (1978)
6. P. Gerlinger, Action of cyclophospyamide injected into the mother on the development of the shape of the body of rabbit embryos. C R Seances Soc. Biol. Fil. **158**, 2154–2157 (1964)
7. S. Chaube, W. Kreis, K. Uchida, M.L. Murphy, The teratogenic effect of 1-beta-D-arabinofuranosylcytosine in the rat. Protection by deoxycytidine. Biochem. Pharmacol. **17**(7), 1213–1216 (1968)
8. J.E. Gibson, B.A. Becker, The teratogenicity of cyclophosphamide in mice. Cancer Res. **28**(3), 475–480 (1968)
9. J.M. Trasler, B.F. Hales, B. Robaire, Chronic low dose cyclophosphamide treatment of adult male rats: effect on fertility, pregnancy outcome and progeny. Biol. Reprod. **34**(2), 275–283 (1986)
10. D.J. Merchant, Prostatic tissue cell growth and assessment. Semin. Oncol. **3**(2), 131–140 (1976)
11. Y. Clermont, B. Perey, Quantitative study of the cell population of these miniferous tubules in immature rats. Am. J. Anat. **100**(2), 241–267 (1957)
12. R.E. Mancini, O. Vilar, J.C. Lavieri, J.A. Andrada, J.J. Heinrich, Cytological and cytochemical study of the development of the Leydig cells in the normal human tasticle. Rev. Soc. Argent Biol. **36**, 443 (1960)
13. C. Huckins, Y. Clermont, Evolution of gonocytes in the rat testis during lateembryonic and early post-natal life. Arch. Anat. Histol. Embryol. **51**(1), 341–354 (1968)
14. L.J. Pelliniemi, P.L. Kellokumpu-Lehtinen, M. Dym, Development of the testes. Duodecim **95**(16), 940–946 (1979)
15. B. Gondos, L.A. Conner, Ultrastructure of developing germ cells in the fetalrabbit testis. Am. J. Anat. **136**(1), 23–42 (1973)
16. S.S. Guraya, Recent progress in the morphology, histochemistry, biochemistry, and physiology of developing and maturing mammalian testis. Int. Rev. Cytol. **62**, 187–309 (1980)
17. D. Brittinger, The mutagenic effect of end oxan on the mouse. Hum. Genetik. **3**(2), 156–165 (1966)
18. R.H. Heller, H.W. Jones Jr., Production of ovarian dysgenesis in the rat and human by busulphan. Am. J. Obstet. Gynecol. **1**(89), 414–420 (1964)
19. K.A. Rich, D.M. De Kretser, Effect of differing degrees of destruction of the rat *Seminiferous epithelium* on levels of serum follicle stimulating hormone and androgen binding protein. Endocrinology **101**(3), 959–968 (1977)
20. P.M. Krueger, G.D. Hodgen, R.J. Sherins, New evidence for the role of the Sertolicell and spermatogonia in feedback control of FSH secretion in male rats. Endocrinology **95**(4), 955–962 (1974)
21. P. Dierickx, G. Verhoeven, Effect of different methods of germinal cell destruction on rat testis. J. Reprod. Fertil. **59**(1), 5–9 (1980)
22. E.M. Eddy, Fine structural observations on the form and distribution of nuage ingerm cells of the rat. Anat. Rec. **178**(4), 731–757 (1974)
23. Y. Clermont, A. Mauger, Existence of a spermatogonialchalone in the rat testis. Cell Tissue Kinet. **7**(2), 165–172 (1974)
24. H.M. Beaumont, Changes in the radiosensitivity of the testis during foetaldevelopment. Int. J. Radiat. Biol. Relat. Stud. Phys. Chem. Med. **2**, 247–256 (1960)

25. G. Hughes, Radiosensitivity of male germ-cells in neonatal rats. Int. J. Radiat. Biol. Relat. Stud. Phys. Chem. Med. **4**, 511–519 (1962)
26. Y. Yasuda, T. Kihara, T. Tanimura, H. Nishimura, Gonadal dysgenesis induced by prenatal exposure to ethinyl estradiol in mice. Teratology **32**(2), 219–227 (1985)
27. M.T. Hochereau-de Reviers, Variation in the stock of testicular stem cells and in the yield of spermatogonial divisions in ram and bull testes. Andrologia **8**(2), 137–146 (1976)
28. B. Gondos, D.C. Paup, J. Ross, R.A. Gorski, Ultrastructural differentiation of Leydig cells in the fetal and postnatal hamster testis. Anat. Rec. **178**(3), 551–565 (1974)
29. P. Collins, D. Lacy, Studies on the structure and function of the mammaliantestis. IV. Steroid metabolism in vitro by isolated interstitium and seminiferoustubules of rat testis after heat sterilization. Proc. R. Soc. Lond. B Biol. Sci. **186**(1082), 37–51 (1974)
30. M. Partington, B.W. Fox, H. Jackson, Comparative action of some methane sulphonic esters on the cell population of the rat testis. Exp. Cell Res. **33**, 78–88 (1964)
31. A. Zschauer, C. Hodel, Drug-induced histological changes in rat seminiferoustubular epithelium. Arch. Toxicol. Suppl. **4**, 466–470 (1980)
32. P. Propping, G. Röhrborn, W. Buselmaier, Comparative investigations on the chemical induction of point mutations and dominant lethal mutations in mice. Mol. Gen. Genet. **117**(3), 197–209 (1972)
33. J.A. Botta Jr., H.C. Hawkins, J.H. Weikel Jr., Effects of cyclophosphamide on fertility and general reproductive performance of rats. Toxicol. Appl. Pharmacol. **27**(3), 602–611 (1974)
34. W. Hilscher, B. Hilscher, Kinetics of the male gametogenesis. Andrologia **8**(2), 105–116 (1976)
35. E. Steinberger, M. Ficher, Differentiation of steroid biosynthetic pathways in developing testes. Biol. Reprod. **1**(Suppl 1), 119–133 (1969)
36. W.B. Neaves, Gonadotropin-induced proliferation of endoplasmic reticulum in an androgenic tumor and its relation to elevated plasma testosterone levels. Cancer Res. **35**(10), 2663–2669 (1975)
37. E. Roosen-Runge, D. Anderson, The development of the interstitial cells in the testis of the albino rat. Acta Anat. (Basel) **37**, 125–137 (1959)
38. G.J. Bloch, J. Masken, C.L. Kragt, W.F. Ganong, Effect of testosterone on plasma LH in male rats of various ages. Endocrinology **94**(4), 947–951 (1974)
39. K.J. Reddy, D.J. Svoboda, Lysosomal activity in sertoli cells of normal and degenerating seminiferous epithelium of rat testes: an ultrastructural and biochemical study. Am. J. Pathol. **51**(1), 1–17 (1967)

Chapter 5
Effect of Arsenic Exposure in Reproductive Health

Arsenic contamination is a global health problem. A large number of people are exposed to arsenic mostly through the contaminated water, soil, food, etc., around the world by inhalation and ingestion and through drinking contaminated water. There are reports which indicated that a number of countries have severe groundwater arsenic contamination problem which deteriorating the health of the affecting population that depend upon the concentration of the arsenic in groundwater.

Both epidemiological and experimental studies indicate that arsenic exposure may have adverse effect on both male and female reproduction and pregnancy outcome. The arsenic exposure may also affect the placental vasculogenesis which may have role in impairment in pregnancy or its outcome. The arsenic exposures are also reported to affect semen quality and hormonal homeostasis as well as induce oxidative stress in both sexes that may be associated with arsenic induces adverse reproductive health. There is an urgent need to make necessary provision for clean drinking water in the area where groundwater is contaminated with arsenic in order to protect human health including reproductive health.

Arsenic (As) is the naturally occurring element in the earth's crust and highly toxic in its inorganic trivalent form as compared to organic forms. Globally, a substantial number of people are exposed to high levels of inorganic arsenic generally through drinking of contaminated water, eating of foods which are grown by arsenic-contaminated water, some industrial processes, eating contaminated food and smoking tobacco. Arsenic contamination in the drinking water is considered as one of the serious worldwide health threats. A population of more than 100 million people worldwide are at risk [1]. It is estimated that about 20 and 45 million people in Bangladesh alone are at risk of being exposed to arsenic concentrations that are more than the national standard of Bangladesh, i.e., 50 µg/l and the WHO guideline of 10 µg/l, respectively [2].

Arsenic is a metalloid, extremely toxic and carcinogenic, and extensively available in the form of sulfides or oxides or as a salt of sodium, iron, calcium, copper, etc. Arsenic occurs in two forms (inorganic and organic). Inorganic arsenic compounds (found in water) are highly toxic, while organic arsenic compounds (found in seafood)

A. K. Saxena and A. Kumar, *Fish Analysis for Drug and Chemicals Mediated Cellular Toxicity*, SpringerBriefs in Applied Sciences and Technology, https://doi.org/10.1007/978-981-15-4700-3_5

are less harmful to health [3]. Calamity of arsenic toxicity has already been reported from many countries, i.e., Bangladesh, India, Nepal, Cambodia, Myanmar, Taiwan, Mongolia, Vietnam, Pakistan, China, Afghanistan, Argentina, Mexico, Chile and America [4, 5]. Asia is the highly affected region due to arsenic toxicity in the world. In India, highest distressed zones due to arsenic toxicity are West Bengal, Chhattisgarh, Bihar, Jharkhand, Uttar Pradesh, Andhra Pradesh and even some states of northeastern region of India also.

Long-term arsenic exposure from drinking water and food can cause skin lesions and cancer. It has also been related to developmental effects, neurotoxicity, cardiovascular disease and diabetes [3]. Further, arsenic exposure is also reported to be associated with adverse pregnancy outcomes and infant mortality, with effects on child health, etc. [6]. Recently, Kim and Kim [7] stated that arsenic toxicity depends on dose, route and gestation periods of exposure. In males, inorganic arsenic causes reproductive impairments including reductions of the accessory sex organ weights, testis weights and epididymal sperm counts. Further, inorganic arsenic exposure also induces impairments of spermatogenesis, reductions of testosterone and gonadotropins, and disruptions of steroidogenesis. They also reported that prenatal exposure to inorganic arsenic causes adverse pregnancy outcomes and children's health impairments and induces premature delivery, spontaneous abortion and stillbirth. The data on human and relevant animal data on reproduction with respect to arsenic exposure are summarized in this communication.

5.1 Experimental Design

The literature was collected through searching various Web sites such as PubMed, Google, and TOXNET, and also consulted relevant books and journals pertaining to reproductive, environmental and occupational health with respect to arsenic. The data is summarized in Table 5.1 with respect to male and Table 5.2 with regard to female reproductions and arsenic exposure. The data pertaining to compounds tested against the toxicity of arsenic is presented in Table 5.3.

5.1.1 Arsenic and Male Reproduction

There are some reports of adverse effect of arsenic on human male reproduction, and a reasonable experimental data also existed. Hseish et al. (2008) revealed that chronic arsenic exposure has a negative impact on erectile dysfunction (ED). They reported that subjects with arsenic exposure (>50 ppb) had a higher risk of developing ED. Later, Mathur et al. [8] reported that the management of infertility problems has become an increasingly essential part of health services. The large number of couples seeks fertility treatment due to poor semen quality, and there is evidence that male reproductive function seems to have deteriorated considerably due to the presence

Table 5.1 Arsenic exposure and male reproduction

S. no.	Exposure	Effects	Reference
Epidemiological studies			
1	Chronic As exposure	Erectile dysfunction in men	Hseish et al. (2008)
2	Inorganic arsenic via staple diet rice	Deterioration of semen quality	Xu et al. [10]
3	Groundwater contamination with the heavy metals like arsenic and cadmium	Altered semenological parameters, lower expression of accessory sex gland markers (fructose, acid phosphatase, neutral α-glucosidase) and alteration in sperm functional parameters (hypo-osmotic swelling, acrosome reaction, nuclear chromatin de-condensation)	Sengupta et al. [9]
4	Environmental As exposure	Infertility, oxidative stress and sexual hormone disruption in males Increased risk of lower sperm motility. Lower semen volume and LH levels	Shen et al. [12] Meeker et al. [14, 15]
5	Environmental As exposure and UMI	Significantly higher concentration of a different moiety of arsenic in the cases which indicated that low-level environmental arsenic exposure positively associated with UMI risk	Wang et al. [16]
Experimental studies			
1	As exposure through drinking water	Decrease in sperm count, motility and the elevation in percentage of abnormal sperm in mice	Pant et al. [18]
2	As_2O_3 exposure orally	Changes androgenic activity with reduced accumulation of spermatozoa, imbalance hormonal level and changes in sperm count and motility in mice	Ali et al. [17]

(continued)

Table 5.1 (continued)

S. no.	Exposure	Effects	Reference
3	High As exposure	Suppresses the gonadotropin-releasing hormone as well as gonadotropin secretion. Reduction in sperm number, viability and motility. Massive degeneration of germ cells and alterations in the level of LH, FSH and testosterone	Zubair et al. [21]
4	As exposure through drinking water	Decline in relative testicular weight and in seminiferous tubular diameter in mice	Sanghamitra et al. [22]
5	As (4, 5 or 6 mg/kgbwt) (ip)	Suppresses spermatogenesis, gonadotropin and testosterone release	Sarkar et al. [19]
6	As (5 mg/kgbwt) through drinking water	Germ cell degeneration, inhibits androgen production, affects the pituitary gonadotropins	Jana et al. [20]
7	As exposure through water(20 or 40 mg/l)	Decreased epididymal sperm counts, glutathione and elevation in protein carbonyl levels in mice	Chang et al. [26]

of considerable amount of arsenic in drinking water in some parts of the world. Sengupta et al. [9] found that a correlation existed between altered semenological parameters, lower expression of accessory sex gland markers (fructose, acid phosphatase, neutral α-glucosidase) in the seminal plasma and significant differences of the sperm functional parameters (hypo-osmotic swelling, acrosome reaction, nuclear chromatin de-condensation) in subjects exposed with groundwater contamination with the heavy metals like arsenic and cadmium in Southern Assam, India. Xu et al. [10] also reported deterioration of semen quality with exposure to arsenic in reproductive age group. They mentioned that in China and throughout Asia, huge populations depend on rice as a staple food, which can concentrate higher levels of inorganic arsenic than wheat and hence rice may be the leading source of arsenic intake. They reported that the arsenic exposure may be associated with reduced human semen quality. Very recently, Bakhat et al. [11] also mentioned that among food crops, rice contains the highest concentration of arsenic. Thus, the general population may be exposed to inorganic arsenic via rice.

Shen et al. [12] showed that elevated urinary concentrations of inorganic arsenic are significantly related to infertility in male, and arsenic species may exert toxicity via disrupting mechanisms of sexual hormone and oxidative stress, as indicated by related biomarkers. Wirth and Minjal [13] also reported the effects of low-level arsenic exposure on human male reproductive outcomes. A cross-sectional study

Table 5.2 Arsenic exposure and female reproduction and outcome

S. no.	Exposure	Effects	Reference
Epidemiological studies			
1	As exposure	Delayed menarche	Sengupta [39] Sen and Chaudhuri [40]
2	As through drinking water in reproductive-age women	Higher stillbirths and miscarriages Spontaneous abortion, stillbirth and neonatal death High prenatal and neonatal mortality Pregnancy may be associated with increased risk of neonatal death Risks of fetal loss and infant death Impaired fetal, infant health. Increases oxidative stress, inflammation in the placenta Both maternal DNA damage and adverse newborn health	Sen and Chaudhuri [41] Milton et al. [45] Ahmad et al. [42] Rudnai and Gulyas [44] Hopenhayn et al. [52] Myers et al. [60] Rahman et al. [59] Ahmed et al. [68] Chou et al. [71]
3	As exposure through drinking water and its side effects on pregnancy outcome	Reduction in birth weight Reduction in birth weight appears to be mediated through decreasing gestational age and by lower maternal gain	Hopenhayn et al. [48], Yang et al. [49], Huyck et al. [51] Kile et al. [50]
4	Environmental As exposures	Increases the risk of stillbirth	Ihrig et al. [43]
5	High concentrations of arsenic through drinking water	Sixfold increased risk of stillbirth	Ehrenstein et al. [46]
6	Chronic As exposure through drinking water in pregnant women	Birth defects, other outcomes (stillbirth, low birth weight, childhood stunting)	Kwok et al. [47]
7	Women exposed to arsenic during smelter	Infants of lower birth weight, spontaneous abortions and congenital malformations	Nordstrom et al. [53–55]
8	Pregnant women exposed to copper smelter	As concentration is higher in placenta higher risk of oxidative damage	Tabacova et al. [56]
9	Acute high doses of arsenic in food during prenatal exposure	Intrauterine fetal death Miscarriage and early neonatal death	Bolliger et al. [57], Lugo et al. [58]

(continued)

Table 5.2 (continued)

S. no.	Exposure	Effects	Reference
10	Low-dose As exposure	Spontaneous abortion, stillbirth, developmental impairment and congenital malformation	Beckman and Nordstrom [61], Zierler et al. [62], Aaschengrau et al. [63], Börzsönyi et al. [64]
Experimental studies			
1	As exposure (4 mg/kg) to goats for 7 weeks, then 5 mg/kgbwt for next 8 weeks orally	Thickened myometrial layer and shortened mucosal folds in uterine tube. Reduction in the number and size of endometrial glands	Islam et al. [77]
2	As exposure drinking water	Suppresses ovarian steroidogenesis, prolongs diestrus and degenerates ovarian follicular and uterine cells in rats. Increases meiotic aberrations in oocytes. Decreases cleavage and preimplantation development	Chattopadhyay et al. [78], Zhang et al. [80] Navarro et al. [79]
3	Inorganic arsenic exposure in (hamsters, mice, rats, rabbits)	Induces developmental toxicity, including malformation, death and growth retardation	Golub et al. [81]
4	As (10 or 12 mg/kg) (ip)	Embryo lethal; maternal deaths in pregnant mice	Hood [82]
5	As$_2$O$_3$ orally prior 14 days (10 mg/kg/day) to mating and until gestational day 19	Decreased food consumption, body gain, increased liver, kidney wt. and stomach aberrations	Holson et al. [83]
6	As (50, 100 and 200 ppm) through drinking water to immature rats	Adverse effect of uterine function and structure. Decrease in uterine wt. and length	Akram et al. [84]
7	As exposure for 10 weeks	Alteration in levels of estradiol, FSH, LH and prolactin	Zhang and Tang (2005)
8	As exposure orally (4 μg/ml) through water	Disrupted the levels of gonadotropins and estradiol, degeneration of luminal epithelial, stromal and myometrial cells of the uterus	Chatterjee and Chatterji [85]

(continued)

Table 5.2 (continued)

S. no.	Exposure	Effects	Reference
9	As-contaminated water to rats for seven estrous cycles	Diminished ovarian key steroidogenic enzyme activities, gonadotropins and estradiol signaling, disrupted ovarian and uterine growth	Chattopadhyay and Ghosh [86]
10	As drinking water (0.4 ppm) to rats for 16 and 28 days	Reduction in LH, FSH, estrogen, activities of ovarian delta 5-3 β-HSD and 17 β-HSD, diminution in the wt. of ovary, uterus and vagina	Chattopadhyay et al. [78]
11	As (10 ppb) through drinking water to dam	Growth deficits in the offspring and cross-fostering reversed the deficit. Arsenic-exposed dams displayed altered liver and breast milk triglyceride levels	Kozul-Horvath et al. [90]
12	As through drinking water	Spontaneous abortion by virtue of aberrant placental vasculogenesis in mice	He et al. [92]
13	As to offspring during postnatally	Suppresses insulin-like growth factor 1 (IGF-1) resulting in delayed sexual maturation	Reilly et al. [95]

of Meeker et al. [14] on men attending infertility clinics in Michigan, USA, also found a significantly increased risk of lower sperm motility with exposure to environmental levels of arsenic. In another report by the same authors, increasing arsenic level was associated with increasing odds for low luteinizing hormone (LH) levels, after adjusting for age, BMI and current smoking [15]. Very recently, Wang et al. [16] explored the relationship between non-geogenic environmental arsenic exposure and risk of unexplained male infertility (UMI). Infertile men with normal semen as cases and fertile men as controls were recruited. Five urinary arsenic species, i.e., pentavalent arsenate (As_i^V), trivalent arsenite (As_i^{III}), methylated to monomethylarsonic acid (MMA^V), dimethylarsinic acid (DMA^V) and arsenobetaine (AsB), were determined in fertile and infertile men. They found that concentrations of As_i^V, AsB, MMA^V, DMA^V, total inorganic arsenic and total arsenic were significantly higher in the cases which indicate that exposure to low-level environmental arsenic is associated with UMI risk.

Several experimental studies on effect of arsenic on various aspects of male reproductive system are also available. Ali et al. [17] observed alterations in androgenic activity with reduced accumulation of spermatozoa and imbalance hormonal level in male mice with arsenic exposure and a significant change in sperm count and motility. Pant et al. [18] recorded the decrease in sperm count and motility, and

Table 5.3 Remedial compounds against arsenic toxicity

S. no.	Exposure	Protective compound	Remedial effect	Reference
Male reproduction				
1	As (5 mg/kg) intragastrically	Melatonin (25 mg/kg)	Rats' testicular injury ameliorated by melatonin	Uygur et al. [27]
2	As (100 ppm) drinking water	Vitamin E (400 mg/kg)	Reversed oxidative stress in testis of rats	Sudha and Matanghi [25]
3	As (10 mg/kg) orally in mice	Arjunolic acid (20 mg/kgbwt)	Prevents testicular oxidative stress and injury	Manna et al. [24]
4	As (10 mg/kg) orally	Coenzyme Q10 (10 mg/kg)	Elevates serum testosterone level, suppressed LPO, antioxidant system defenses, mitigate the elevation of TNF-α and nitric oxide	Fouad et al. [28]
		Thymoquinone (10 mg/kg)	Ameliorates the declines of serum testosterone, reduced glutathione level in testis	Fouad et al. [29]
5	As (6.3 mg/kg) orally	*Spirulina platensis* (300 mg/Kg)	Prevent arsenic-induced testicular oxidative damage in rats	Bashandy et al. [30]
6	As (100 mg/l) drinking water	Polydatin (50, 100 and 200 mg/kg)	Reduced LPO, enhances antioxidant defense mechanism and regenerates tissue damage in testis	Ince et al. [31]
7	As (8 mg/kg/day) orally	Vitamin E (100 mg/kg/day)	Ameliorates the adverse effects on sperm number and diameters of tubule in rats	Momeni and Eskandari [33]
8	As (5 mg/kg) (ip)	Curcumin (100 mg/kg)	Ameliorates sperm quality parameters in rats	Momeni and Eskandari [34]

(continued)

Table 5.3 (continued)

S. no.	Exposure	Protective compound	Remedial effect	Reference
9	As (10 mg/kgbwt/day)	Ginger extract (500 mg/kgbwt/day)	Attenuate the decrease in sperm functions, enhance reproductive hormones level and reverse the oxidative stress	Morakinyo et al. [35]
10	As (3 mg/kgbwt/day) orally	High-protein diet	Effective against toxic effect of arsenic on male gonad of rat	Mukherjee and Mukhopadhyay [32]
11	As (50 ppm)	Quercetin (50 mg/kg)	Reversal of oxidative stress in testis	Jahan et al. [37]
Female reproduction				
12	As (4 ppm) drinking water	Catechin	Catechin increases sexual hormones in rats	Hedayati et al. [98]
13	As (3 ppm)	HPD	HPD may have role in the protection of reproductive damage caused by arsenic	Mondal et al. [99]
14	As (35 mg/kg) (ip) on 8th gestation day	Vitamins C and E	Vitamins C and E upturn the reduction in maternal wt	Qureshi and Tahir (2013)
15	As (0.4 ppm/100 g) drinking water	Sodium selenite	Increases ovarian steroidogenic enzymes, plasma levels of LH, FSH, estradiol as well as ovarian and uterine peroxidase	Chattopadhyay et al. [87]

the percentage of morphological abnormal sperm was elevated in arsenic intoxicated mice. Additionally, the activities of marker testicular enzymes also were altered. Earlier, Sarkar et al. [19] also reported that arsenite has a suppressive influence on spermatogenesis and gonadotropin and testosterone release in rats. Later, Jana et al. [20] found that arsenic causes testicular toxicity by germ cell degeneration and impedes androgen production in adult rats maybe by affecting pituitary gonadotropins. In recent year, Zubair et al. [21] reviewed that high arsenic exposure may suppress the sensitivity of gonadotroph cells to gonadotropin-releasing hormone (GnRH) as well as gonadotropin secretion by elevating plasma levels of glucocorticoids. These alterations lead to the gonadal toxicity in animals and cause reduction in sperm number, viability and motility. Massive degeneration of germ cells and alterations in the level of LH, follicular stimulating hormone (FSH) and testosterone were also observed. Sanghamitra et al. [22] studied the effect of arsenic on the testicular tissue of mice and found a significant decline in the relative testicular weight, decline in seminiferous tubular diameter, various gametogenic cell population, i.e., resting spermatocyte, pachytene spermatocyte and step-7-spermatid except spermatogonia and leydig cell atrophy was also observed. Later, Chiou et al. [23] concluded that arsenic trioxide (As_2O_3) treatment deteriorated sperm mobility and viability and disturbed spermatogenesis via decreasing gene expression of the key enzymes which involved in testosterone synthesis.

A number of natural and synthetic compounds were tested against the reproductive toxicity of arsenic in different animal species by various investigators. Manna et al. [24] suggested that an impaired antioxidant defense mechanism followed by oxidative stress is the major cause of arsenic-induced toxicity. In addition, arsenic intoxication enhanced testicular arsenic content, lipid peroxidation (LPO), protein carbonylation and the level of glutathione disulfide. They found that pretreatment with arjunolic acid reverses testicular oxidative stress and injury to the histological structures of the testes and this may be due to its intrinsic antioxidant property. Later, Sudha and Matanghi [25] also reported a significant increase in the level of LPO and decrease in the levels of antioxidants and enzyme activities were observed in arsenic exposed rats. Co-administration of α-tocopherol reversed the hostile effect of arsenic. Earlier, Chang et al. [26] also reported that oxidative stress to be a major cause of male reproductive failure. They observed that ascorbic acid ameliorates the arsenic-induced decline in epididymal sperm counts and testicular weights and decreased glutathione levels and elevated levels of protein carbonyl content. Later, Uygur et al. [27] studied the protective effects of melatonin against arsenic-induced apoptosis and oxidative stress in rat testes. They found that testicular injury induced by arsenic was ameliorated by the administration of melatonin. The number of apoptotic germ cell was increased, and the number of proliferating cell nuclear antigen-positive germ cell was decreased in testis after arsenic administration. The decreased SOD, catalase and glutathione peroxidase activities as well as increased malondialdehyde (MDA) levels in testis were counteracted by melatonin.

In addition to prevention of arsenic-induced toxicity by ascorbic acid, vitamin E and melatonin, Fouad et al. [28] reported that coenzyme Q10 significantly increased serum testosterone level, suppressed LPO, restored the depleted antioxidant defenses and attenuated the increases of tumor necrosis factor-α and nitric oxide resulted from arsenic administration. Later, they also tested the protective effect of thymoquinone (TQ) against sodium arsenite-induced testicular injury. TQ significantly attenuated the arsenic-induced decreases of serum testosterone, reduced glutathione level and decrease in elevations of MDA in testicular tissue and nitric oxide levels. Further, TQ decreases the expression of caspase-3 and inducible nitric oxide synthase in testicular tissue [29]. Very recently, Bashandy et al. [30] examined the remedial role of *Spirulina platensis* (*S. platensis*) against arsenic toxicity. Arsenic caused a significant arsenic accumulation in testicular tissues and diminution in the levels of

SOD, catalase, reduced glutathione in testicular tissue, zinc and plasma testosterone, LH, triiodothyronine (T3) and thyroxine (T4) levels, sperm motility and count and arsenic led to a significant increase in testicular MDA, tumor necrosis factor alpha, nitric oxide and sperm abnormalities. They found that *S. platensis* prevent these alterations. Recently, Ince et al. [31] studied protective effects of polydatin (PD) against reproductive effects of chronic arsenic exposure in rats. The results reveal that PD decreases arsenic-induced LPO, enhances the antioxidant defense mechanism and regenerates tissue damage in testis. Mukherjee and Mukhopadhyay [32] found that As_2O_3 caused an increase in seminiferous tubular luminal size together with declined accumulation of spermatozoa, signs of necrotic changes with disarray in cellular organization and decline in sperm count, viability and motility. They also reported that arsenic toxicity on male gonad may be counteracting by high-protein diet supplementation.

Momeni and Eskandari [33] investigated the hostile effects of sodium arsenite on the male reproductive system in rats and its modification by vitamin E. Vitamin E ameliorated toxic effects of sodium arsenite on sperm number as well as the diameters of tubule and lumen. Later, the same authors mentioned that curcumin also ameliorated the toxic effect of sodium arsenite on a number of sperm parameters in adult mice [34]. Earlier, Morakinyo et al. [35] studied the effect of sodium arsenite with or without aqueous ginger extract (500 mg/kgbwt/day) in rats. Aqueous ginger extract was found to be beneficial against adverse change induced by arsenite in the reproductive organ weight, mitigate the decrease in sperm functions and increase plasma reproductive hormone level along with enhanced antioxidant activities and reduced peroxidation.

Recently, Eskandari and Momeni [36] investigated the curative effects of sily-marin against the adverse effect of arsenic on ram sperm quality. Decreased viability, non-progressive motility and intact mitochondrial membrane potential of the sper-matozoa were found in sodium arsenite treated group, and these could be reversed by silymarin. Jahan et al. [37] studied the remedial effect of quercetin against arsenic-induced reproductive ailments in rats. Arsenic treatment resulted in hostile morpho-logical and histopathological changes in testis including reduced epithelial height and tubular diameter, and increased luminal diameter, and these adverse effects were reduced by quercetin. LPO was significantly suppressed, and depleted antioxidant defense mechanism was restored by the co-treatment of quercetin. Additionally, quercetin treatment resulted in a marked rise in plasma and testicular testosterone concentrations also. Baltaci et al. [38] also reported that quercetin prevents testicular damage induced by arsenic and this may be due to its antioxidant and anti-apoptotic property.

Based on data available, it can be inferred that arsenic affects the male repro-ductive system by declining the sperm count, motility, normal sperm morphology, and alteration in hormone level, testicular injury, induction of reactive oxygen and nitrogen species and caspase-3 in testicular tissue. A number of compounds (mela-tonin, vitamin E, ascorbic acid, coenzyme Q 10, arjunolic acid, thymoquinone, poly-datin, ginger extract, high-protein diet, silymarin, quercetin, etc.) were tested by different investigators to find out suitable compound. Some of these compounds may have promising potential for clinical use after detailed investigations.

5.1.2 Arsenic and Female Reproduction

Females, even pregnant women, are exposed to arsenic generally through contaminated drinking water and ingestion of contaminated food grown in the area where groundwater is contaminated with arsenic. A number of epidemiological reports are available on the exposure to arsenic and female reproduction. Sengupta [39] determines the effect of arsenic exposure on menarcheal age. They found that arsenic-affected female attained menarche at a later age significantly than the other female though both are enjoying similar conditions (economy, education, family size, food habits, geography, etc.) except the consumption of arsenic for long time. Later, Sen and Chaudhuri [40] also found higher mean menarcheal age of 12.5 yrs in women residing in arsenic-affected area as compared to 11.7 yrs in control. These studies point out that arsenic exposure can have a negative impact on menarcheal age. Later, Sen and Chaudhuri [41] studied the effects of arsenic exposure on the pregnancy outcome and found a higher rate of stillbirths and miscarriages than those in the unexposed population. Earlier, Ahmad et al. [42] studied women who were chronically exposed to arsenic through drinking water and found adverse pregnancy outcomes in terms of spontaneous abortion and rates of stillbirth and preterm birth. Earlier, Ihrig et al. [43] assessed environmental arsenic exposures and risk of stillbirth and found statistically significant increase in the risk of stillbirth in the group with the highest exposure to arsenic. Rudnai and Gulyas [44] also reported an increase in spontaneous abortions, stillbirths and prenatal mortality in Karcag, Hungary, due to arsenic exposure through drinking water. However, Milton et al. [45] reported that chronic arsenic exposure through drinking water has the potential to cause adverse pregnancy outcomes, although the association has not been demonstrated conclusively.

Ehrenstein et al. [46] reported that exposure to arsenic at high concentrations (≥ 200 μg/l) was connected with a sixfold elevated risk of stillbirth. Arsenic-related skin lesions were observed in 12 women who had a significantly augmented risk of stillbirth. No relationship was found between arsenic exposure and spontaneous abortion or overall infant mortality. Kwok et al. [47] examined 2006 pregnant women chronically exposed to arsenic through drinking water and found significant relationship between arsenic exposure and birth defects; other outcomes, such as low birth weight, stillbirth, childhood stunting and underweight, were not linked, and relationship may be a statistical anomaly due to the smaller number of high birth defects in the study. However, most of the data suggests that arsenic chronic exposure increases the risk of spontaneous abortion and stillbirth.

Further, Hopenhayn et al. [48] observed that moderate arsenic exposures from drinking water (50 μg/l) during pregnancy are associated with reduction in birth weight. Yang et al. [49] conducted a study to compare the risk of adverse pregnancy outcomes (preterm delivery and birth weight) between an area with high water arsenic levels and area with no arsenic water contamination. They mentioned that arsenic exposure was associated with the risk of preterm delivery (nonsignificantly). The estimated reduction in birth weight was 29.05 g. Recently, Kile et al. [50] also concluded that arsenic exposure during pregnancy was related to lower birth weight and this seems to be facilitated mainly through reducing gestational age and to a lesser extent by reducing maternal weight gain during pregnancy. Earlier, Huyck et al. [51] found that higher maternal hair arsenic measurement in early pregnancy was positively correlated with drinking water arsenic level and reduction in birth weight. The available data provides ample evidence for a possible effect of arsenic exposure through drinking water with the risk of low birth weight and other adverse pregnancy outcome.

Hopenhayn et al. [52] reported high prenatal and neonatal mortality in the mining area of northern Chile in association with arsenic-contaminated water. Earlier, Nordstrom et al. [53–55] also showed that women working in the smelter or living nearby to copper smelter area gave birth to infants of lower birth weight and had a higher incidence of spontaneous abortions and congenital malformations. Later, Tabacova et al. [56] also examined oxidative damage during pregnancy and arsenic exposure from a copper smelter area in Bulgaria. Placental levels of arsenic were highest in areas with the highest environmental contamination and were at higher risk of oxidative damage. Earlier studies also suggested that in humans, prenatal exposure to acute high doses of arsenic has resulted in miscarriage and early neonatal death [57, 58]. Later, Rahman et al. [59] investigated the toxic effect of arsenic on fetal and infant survival. Drinking tube-well water with more than 50 $\mu g/l$ arsenic concentration during pregnancy significantly increased the risks of fetal loss and infant death. Myers et al. [60] also found that drinking water arsenic exposure during pregnancy may be associated with increased risk of neonatal death. Additionally, earlier studies also reported that prolonged low-dose human arsenic exposure has been associated with multiple adverse reproductive outcomes such as spontaneous abortion, stillbirth, developmental impairment and congenital malformation [61–64].

Previously reported that arsenic is related with hostile pregnancy outcomes and infant mortality. However, the explanation of the causal link is hampered by methodological challenges and limited number of studies on dose–response. Earlier, Bloom et al. [65] reviewed epidemiologic studies and concluded that epidemiologic evidence for an increased risk of low birth weight is inadequate although limited evidence for decreases in birth weight and birth size is available. Earlier, they reviewed that maternal exposure to high concentrations of inorganic arsenic in naturally contaminated drinking groundwater has been associated with an increased risk of the spontaneous loss in several epidemiologic studies [66]. Raqib et al. [67] also suggested that in utero arsenic exposure reduced child thymic development and increased morbidity, possibly via immunosuppression. The findings of Ahmed et al. [68] also suggested that the arsenic affected the immune function that may contribute to diminished fetal and infant health. They reported that maternal exposure increases oxidative stress and inflammation in the placenta. Thus, immunosuppression induced by arsenic may also be one of the causative factors behind adverse reproductive outcome.

Earlier, Vahter [69] reviewed arsenic's modes of action which include enzyme inhibition and oxidative stress as well as immune, endocrine and epigenetic effects. He reported that susceptibility to arsenic is dependent on the biomethylation. Methylarsonic acid and dimethylarsinic acid are main metabolites found in urine, and elevated methylarsonic acid is considered as a general risk factor. Arsenic easily passes the placenta, and its effect indicates a moderately increased risk of impaired fetal growth and fetal and infant mortality. Tseng [70] reported that methylation capacity might reduce with increasing dosage of arsenic exposure. Furthermore, women, especially at pregnancy, have better methylation capacity than men, probably due to the effect of estrogen. Chou et al. [71] studied the association between arsenic exposure and oxidative/methylated DNA damage. They reported that maternal urinary inorganic arsenic had positive relation with the methylated N^7-methylguanosine (N^7-MeG) and oxidative 8-oxo-7,8 dihydro 2'deoxyguanosine (8-oxodG) DNA damage biomarkers, and a declined one-minute Apgar score. Maternal inorganic arsenic exposure was related to both maternal DNA damage and adverse newborn health. Earlier, Pilsner et al. [72] also suggested that prenatal exposure of arsenic is related to global DNA methylation in cord blood DNA. Arsenic-induced epigenetic modifications in utero may possibly influence disease outcomes later in life. Further, Chen et al. [73] showed that chronic (>18 wk), low-level (125 ± 500 nM) arsenite exposure induces malignant transformation in normal rat liver cell line. In these arsenic-transformed cells, the cellular S-adenosylmethionine pool was depleted from arsenic metabolism, resulting in global DNA hypomethylation. DNA methylation status in turn may affect the expression of a variety of genes. Later, Reichard and Puga [74] reported that changes in gene methylation status, mediated by arsenic, have been proposed to activate oncogene expression or silence tumor suppressor genes, leading to long-term changes in the activity of genes controlling cell transformation. Intarasunanont et al. [75] also suggested that arsenic carcinogenesis causes epigenetic changes, particularly in DNA methylation. They also reported that in utero arsenic exposure affects DNA methylation, particularly at the p53 promoter region. Later, Koestler et al. [76] found that in utero exposure to low concentration of arsenic may affect the epigenome.

Based on available epidemiological studies, one can infer that arsenic exposure induces adverse effects on pregnancy and its outcome such as premature delivery, spontaneous abortion, developmental impairment, stillbirth, morbidity and prenatal mortality. In addition, arsenic exposure may also affect the female reproductive system by altering hormonal homeostasis.

Experimental studies of Islam et al. [77] suggested that a chronic arsenic exposure might have adverse effects on the female reproductive system of black goat. Earlier studies in female mice and rats showed that inorganic arsenic suppresses ovarian steroidogenesis, prolongs diestrus and degenerates ovarian follicular and uterine cells [78–80]. It also increases meiotic aberrations in oocytes and decreases cleavage and preimplantation development [79]. Arsenic can also induce ovarian and uterine toxicity, and influence neuroendocrine regulation of female sex hormones [78]. Golub et al. [81] also reported that inorganic arsenic can cause developmental toxicity, including malformation, death and growth retardation in four species (hamsters, mice, rats and rabbits) and the developmental toxicity effects are dependent on dose, route and the day of gestation when exposure occurs. When females were dosed chronically for periods that included pregnancy, the primary effect of arsenic on reproduction was a dose-dependent increase in conceptus mortality and postnatal growth retardation. Earlier, Hood [82] reported that sodium arsenite administered ip to pregnant mice at dose levels that were toxic to the dams was embryo lethal; some maternal deaths were also occurred. Later, Holson et al. [83] also reported that administration of arsenic trioxide prior 14 days to mating and continuing until gestational day 19, maternal toxicity was indicated by decreased food consumption and increase in body weight during gestation, increased kidney and liver weights, and stomach abnormalities. No observed adverse effect level dose (NOAEL) was found to be 5 mg/kg/day. Further, they reported that oral As_2O_3 cannot be considered to be a developmental toxicant, nor cause neural tube defects, at maternally toxic dose levels.

Akram et al. [84] examined the toxic effects of arsenic exposure on uterine structure and function, and dose-dependent decrease was observed in uterine weight and length. They concluded that arsenic is a major threat to female reproductive health acting as a reproductive toxicant and as an endocrine disruptor, restricted the structure and function of uterus, by altering the steroid and gonadotropin levels, not only at high dose concentration but at low (50 ppm) levels also. Chatterjee and Chatterji [85] reported that arsenic interrupted the circulating levels of estradiol and gonadotropins, which led to degeneration of luminal epithelial, myometrial and stromal cells of the rat uterus and down-regulate the downstream components of the estrogen signaling pathway. Further, functional and developmental maintenance of the uterus is under the influence of estradiol; thus, structural degeneration induced by arsenic may be attributed to the fall in circulating estradiol levels. Chattopadhyay and Ghosh [86] reported that ingestion of arsenic-contaminated water to rats for seven estrous cycles significantly diminished ovarian key steroidogenic enzyme activities and gonadotropin and estradiol signaling along with ovarian and uterine growth disruption. Earlier, Chattopadhyay et al. [78] mentioned that arsenic affects ovarian steroidogenesis at the dose present in drinking water at wide areas of West Bengal, India. Weights of ovary, uterus and vagina along with biochemical activities of ovarian delta 5-3 β-hydroxysteroid dehydrogenase (delta 5-3 β-HSD) and 17 β-hydroxysteroid dehydrogenase (17 β-HSD) and plasma levels of LH, FSH and estrogen were measured following treatment with sodium arsenite at a dose of 0.4 ppm/rat/day for 16 days (4 estrous cycles) and 28 days (7 estrous cycles) caused reduction in plasma levels of LH, FSH and estrogen along with significant diminution in the activities of ovarian delta 5-3 β-HSD and 17 β-HSD. Later, they reported that dietary supplementation of sodium selenite minimized the gonadal weight loss and increased the activities of the ovarian steroidogenic enzymes as well as the ovarian and uterine peroxidize at the control level and are also able to increase the plasma levels of LH, FSH and estradiol [87].

Recently, Treas et al. [88] evaluated the effect of exposure to arsenic, estrogen and their combination and found variation in the expression of epigenetic regulatory genes and changes in global DNA methylation and histone modification patterns in RWPE-1 cells. These changes were significantly greater in arsenic and estrogen combination treated group than individually. Earlier, Montalbano and Jacobson-Kram [89] reported that arsenic seems to cross the placenta and cord blood levels at parturition being similar to maternal levels. The weight of evidence supports that arsenic is either inactive or extremely weak for the induction of gene mutations in vitro; it is clastogenic and induces sister chromatid exchange (SCE); it does not appear to induce chromosome aberrations in vivo in animals; human beings exposed to arsenic demonstrate higher frequencies of SCE and chromosomal aberrations, and it may affect DNA by the inhibition of DNA repair processes or by its occasional substitution for phosphorous in the DNA backbone.

Kozul-Horvath et al. [90] examined the pups exposed to 10 ppb arsenic, via the dam drinking water. They reported the exposure during the in utero and postnatal period caused in growth deficits in the offspring, which was predominantly a result of diminished nutrients in the dam's breast milk. However, cross-fostering of the pups reversed the growth deficit. Earlier, Waalkes et al. [91] reported that arsenate exposure, though not outright tumorigenic, was associated with proliferative, preneoplastic lesions of the uterus, testes and liver. Estrogen treatment has been related to proliferative lesions and tumors of the female liver, uterus and testes in male. He et al. [92] reported that the placenta is known to utilize vasculogenesis to develop its circulation. They reported that the arsenic exposure causes placental dysmorphogenesis and defective vasculogenesis resulting in placental insufficiency and subsequently leads to spontaneous abortion. The study suggests that arsenic toxicity is causative for mammalian spontaneous abortion by virtue of aberrant placental vasculogenesis. Recently, Patel et al. [93] reported that higher arsenic exposure inhibited the angiogenesis which was dose-dependent in both chorioallantoic membrane assay (CAM assay) and Matrigel assay and altered structural morphology of placenta. Placenta is known to utilize vasculogenesis to develop its vasculature circulation.

Additionally, Ettinger et al. [94] found that arsenic exposure was related to elevated risk of impaired glucose tolerance test at 24–28 weeks gestation and thus arsenic may be linked with increased risk of gestational diabetes. Further, Reilly et al. [95] investigated the effects of prepubertal arsenic exposure on mammary gland development and pubertal onset. They found that prepubertal exposure to arsenite delayed vaginal opening and prepubertal mammary gland maturation. They determined that arsenic accumulates in the liver, disrupts hepatocyte function and suppresses serum levels of the puberty-related hormone insulin-like growth factor 1 (IGF-1). Further, recently Ser et al. [96] reported that maternal arsenic exposure was positively associated with maternal IgG but not IgG cord. Elevated maternal IgG may have implications with regard to maternal morbidity and the placental transfer of specific IgG, and further studies are required to understand how arsenic may affect maternal and child health by modifying the humoral immune system.

Wares et al. [97] have determined effects of arsenic on uterus of female goats. Gross parameters show slight variations in the morphology, size and weight of uterus of arsenic-affected goats. In microscopic level, there were significant variations in arsenic-affected goats. Recently, Hedayati et al. [98] studied the protective role of catechin on sexual hormones of arsenic-intoxicated rats. Rats protected by catechin had higher plasma concentrations of steroidal and gonadotropin hormone. Mondal et al. [99] investigated the protective effect of casein- and pea-supplemented high-protein diet (HPD) in utero ovarian toxicity from arsenic. HPD, by way of antioxidant properties, found to have prospective role in the protection of reproductive damage caused by arsenic. Further, Qureshi and Tahir (2013) reported that women are more susceptible than men to the adverse effects of arsenic, as it interacts with estrogen hormones. The maternal weight gain in mice was reduced in arsenic treated group, and the difference was statistically significant as compared to control and vitamins C and E protected group. They concluded that vitamins C and E are useful in protecting sodium arsenate-induced reduction of weight.

There is adequate data available which indicates that arsenic exposure through drinking water is a major global health problem; it affects both male and female reproduction that might be mediated through the induction of oxidative stress as well as affecting endocrine system directly affecting the reproductive organs, and these effect might be depended upon the arsenic concentration in drinking water or in diet or through environment, and duration of exposure. Female might be more sensitive toward arsenic exposure as it impaired estradiol circulation which might be responsible for development and functional maintenance of the uterus. As earlier, Chattopadhyay et al. [78] reported that arsenic disrupts the ovarian and uterine growth as well as degenerates luminal epithelial, stromal and myometrial cells of uterus as well as arsenic acts as endocrine disruptor also.

Acknowledgments This chapter is contributed by Dr. Sunil Kumar, Ph.D., Scientist, NIOH Ahmadabad, Gujrat, India.

References

1. P. Ravenscroft, H. Brammer, K. Richards, *Arsenic pollution: a global synthesis* (Wiley, West Sussex, 2009). https://doi.org/10.1002/9781444308785
2. S.V. Flanagan, R.B. Johnston, Y. Zheng, Arsenic in tube well water in Bangladesh: health and economic impacts and implications for arsenic mitigation. Bull. World Health Organ. **90**, 839–846 (2012)
3. WHO, *Arsenic Fact Sheet 2016* (2016), http://www.who.int/mediacentre/factsheets/fs372/en/
4. K.F. Akter, G. Owens, D.E. Davey, R. Naidu, Arsenic speciation and toxicity in biological systems. Rev. Environ. Contamin. Toxicol. **184**, 97–149 (2005)
5. J.D. Ayotte, D.L. Montgomery, S.M. Flanagan, K.W. Robinson, Arsenic in groundwater in eastern New England: occurrence, controls, and human health implications. Environ. Sci. Technol. **37**, 2075–2083 (2003)
6. R. Quansah, F.A. Armah, D.K. Essumang, I. Luginaah, E. Clarke, K. Marfoh, S.J. Cobbina, E. Nketiah-Amponsah, P.B. Namujju, S. Obiri, M. Dzodzomenyo, Association of arsenic with adverse pregnancy outcomes/infant mortality: a systematic review and meta-analysis. Environ. Health Perspect. **123**, 412–421 (2015)
7. Y.J. Kim, J.M. Kim, Arsenic toxicity in male reproduction and development. Dev. Reprod. **19**, 167–180 (2015)

8. N. Mathur, G. Pandey, G.C. Jain, Male reproductive toxicity of some selected metals: a review. J. Biol. Sci. **10**, 396–404 (2010)
9. M. Sengupta, I. Deb, G.D. Sharma, K.K. Kar, Human sperm and other seminal constituents in male infertile patients from arsenic and cadmium rich areas of Southern Assam. Syst. Biol. Reprod. Med. **59**(4), 199–209 (2013)
10. W. Xu, H. Bao, F. Liu, L. Liu, Y.G. Zhu, J. She, S. Dong, M. Cai, L. Li, C. Li, H. Shen, Environmental exposure to arsenic may reduce human semen quality: associations derived from a Chinese cross-sectional study. Environ. Health **11**, 46 (2012)
11. H.F. Bakhat, Z. Zia, S. Fahad, S. Abbas, H.M. Hammond, A.N. Shahzad, F. Abbas, H. Alhorby, H. Sahid, Arsenic uptake, accumulation and toxicity in rice plants: possible remedies for its detoxification: a review. Environ. Sci. Pollut. Res. **24**, 9142–9158 (2017)
12. H. Shen, W. Xu, J. Zhang, M. Chen, F.L. Martin, Y. Xia, L. Liu, S. Dong, Y.G. Zhu, Urinary metabolic biomarkers link oxidative stress indicators associated with general arsenic exposure to male infertility in a Han Chinese population. Environ. Sci. Technol. **47**(15), 8843–8851 (2013)
13. J.J. Wirth, R.S. Minjal, Adverse effects of low level heavy metal exposure on male reproductive function. Syst. Biol. Reprod. Med. **56**, 147–167 (2010)
14. J.D. Meeker, M.G. Rossano, B. Protas, M.P. Diamond, E. Puscheck, D. Daly, N. Paneth, J.J. Wirth, Cadmium, lead, and other metals in relation to semen quality: human evidence for molybdenum as a male reproductive toxicant. Environ. Health Perspect. **116**, 1473–1479 (2008)
15. J.D. Meeker, M. Rossano, B.M. Protas, V. Padmanahban, M.P. Diamond, E. Puscheck, D. Daly, N. Paneth, J.J. Wirth, Environmental exposure to metal and male reproductive hormones: circulating testosterone is inversely associated with blood molybdenum. Fertil. Steril. **93**, 130–140 (2008)
16. X. Wang, J. Zhang, X. Xu, Q. Huang, L. Liu, M. Tian, Y. Xia, W. Zhang, H. Shen, Low-level environmental arsenic exposure correlates with unexplained male infertility risk. Sci. Total Environ. **571**(15), 307–313 (2016)
17. M. Ali, S.A. Khan, P. Dubey, A. Nath, J.K. Singh, R. Kumar, A. Kumar, Impact of arsenic on testosterone synthesis pathway and sperm production in mice. Innov. J. Med. Health Sci. **3**, 185–189 (2013)
18. N. Pant, R.C. Murthy, S.P. Srivastava, Male reproductive toxicity of sodium arsenite in mice. Hum. Exp. Toxicol. **23**, 399–403 (2004)
19. M. Sarkar, G. Ray Chaudhuri, A. Chattopadhyay, N.M. Biswas, Effect of sodium arsenite on spermatogenesis, plasma gonadotrophins and testosterone in rats. Asian J. Androl. **1**, 27–31 (2003)
20. K. Jana, S. Jana, P.K. Samanta, Effects of chronic exposure to sodium arsenite on hypothalamo-pituitary-testicular activities in adult rats: possible an estrogenic mode of action. Reprod. Biol. Endocrinol. **4**, 9 (2006)
21. M. Zubair, M. Ahmad, Z.I. Qureshi, Review on arsenic-induced toxicity in male reproductive system and its amelioration. Andrologia **49**, e12791 (2017)
22. S. Sanghamitra, J. Hazra, S.N. Upadhyay, R.K. Singh, R.C. Amal, Arsenic induced toxicity on testicular tissue of mice. Indian J. Physiol. Pharmacol. **52**(1), 84–90 (2008)
23. T.J. Chiou, S.T. Chu, W.F. Tzeng, Y.C. Huang, C.J. Liao, Arsenic trioxide impairs spermato-genesis via reducing gene expression levels in testosterone synthesis pathway. Chem. Res. Toxicol. **21**, 1562–1569 (2008)
24. P. Manna, M. Sinha, P.C. Sil, Protection of arsenic-induced testicular oxidative stress by arjunolic acid. Redox Rep. **3**(2), 67–77 (2008)
25. K. Sudha, S.K. Matanghi, Reversing effect of α-tocopherol in arsenic induced toxicity in albino rats. Int. J. Pharm. Pharm. Sci. **4**(5), 282–284 (2012)
26. S.I. Chang, B. Jin, P. Youn, C. Park, J.D. Park, D.Y. Ryu, Arsenic-induced toxicity and the protective role of ascorbic acid in mouse testis. Toxicol. Appl. Pharmacol. **218**(2), 196–203 (2007)

27. R. Uygur, C. Aktas, V. Caglar, E. Uygur, H. Erdogan, O.A. Ozen, Protective effects of melatonin against arsenic-induced apoptosis and oxidative stress in rat testes. Toxicol. Ind. Health 1–12 (2013)
28. A.A. Fouad, Sultan Al, M.T. Yacoubi, Coenzyme Q10 counteracts testicular injury induced by sodium arsenite in rats. Eur. J. Pharmacol. **655**(1–3), 91–98 (2011)
29. A.A. Fouad, W.H. Albulai, I. Jresat, Protective effect of thymoquinone against arsenic-induced testicular toxicity in rats. Int. J. Med. Health Biomed. Bioeng. Pharm. Eng. **8**(2) (2014)
30. S.A. Bashandy, S.A. El Awdan, H. Ebaid, I.M. Alhazza, Antioxidant potential of *Spirulina platensis* mitigates oxidative stress and reprotoxicity induced by sodium arsenite in male rats. Oxid. Med. Cell. Longev. **7174351**, 1–8 (2016)
31. S. Ince, F. Avdatek, H.H. Demirel, D.A. Acaroz, E. Goksel, I. Kucukkurt, Ameliorative effect of polydatin on oxidative stress mediated testicular damage by chronic arsenic exposure in rats. Andrologia **48**, 518–524 (2016)
32. S. Mukherjee, P. Mukhopadhyay, Studies on arsenic toxicity in male rat gonads and its protection by high dietary protein supplementation. Al Ameen J. Med. Sci. **2**, 73–77 (2009)
33. H.R. Momeni, N. Eskandari, Effect of vitamin E on sperm parameters and DNA integrity in sodium arsenite treated rats. Iran J. Reprod. Med. **10**, 249–256 (2012)
34. H.R. Momeni, N. Eskandari, Curcumin inhibits the adverse effects of sodium arsenite in mouse epididymal sperm. Int. J. Fertil. Steril. **10**(2), 245–252 (2016)
35. A. Morakinyo, P. Achema, O. Adegoke, Effect of *Zingiber officinale* (ginger) on sodium arsenite-induced reproductive toxicity in male rats. Asian J. Med. Res. **13**, 39–45 (2010)
36. F. Eskandari, H.R. Momeni, Protective effect of silymarin on viability, motility and mitochondrial membrane potential of ram sperm treated with sodium arsenite. Int. J. Reprod. Biomed. **14**(6), 397–402 (2016)
37. S. Jahan, N. Iftikhar, H. Ullah, G. Rukh, I. Hussain, Alleviative effect of quercetin on rat testis against arsenic: a histological and biochemical study. Syst. Biol. Reprod. Med. **61**(2), 89–95 (2016)
38. B.B. Baltaci, R. Uygur, V. Caglar, C. Aktas, M. Aydin, O.A. Ozen, Protective effects of quercetin against arsenic-induced testicular damage in rats. Andrologia **48**, 1–12 (2016)
39. M. Sengupta, Does arsenic consumption influence the age at menarche of woman. Indian Pediatr. **41**(9), 960–961 (2004)
40. J. Sen, B.D. Chaudhuri, Effect of arsenic on the onset of menarcheal age. Bull. Environ. Contamin. Toxicol. **79**(3), 293–296 (2007)
41. J. Sen, B.D. Chaudhuri, Arsenic exposure through drinking water and its effect on pregnancy outcome in Bengali women. Arh. Hig. Rada Toksikol. **59**, 271–275 (2008)
42. A. Ahmad, M.H.S.U. Sayed, S. Shampa Barua, M.H. Khan, M.H. Faruquee, A. Jalil, S.A. Hadi, H.K. Talukder, Arsenic in drinking water and pregnancy outcomes. Environ. Health Perspect. **109**, 629–631 (2001)
43. M.M. Ihrig, S.L. Shalat, C. Baynes, A hospital-based case-control study of stillbirths and environmental exposure to arsenic using an atmospheric dispersion model linked to a geographical information system. Epidemiology **9**, 290–294 (1998)
44. P. Rudnai, E. Gulyas, Adverse effects of drinking water-related arsenic exposure on some pregnancy outcomes in Karcag, Hungary. Presented at the 3rd international conference on arsenic exposure and health effects, San Diego, CA, 12–15 July 1998
45. A.H. Milton, W. Smith, B. Rahman, Z. Hasan, U. Kulsum, K. Dear, M. Rakibuddin, A. Ali, Chronic arsenic exposure and adverse pregnancy outcomes in Bangladesh. Epidemiology **16**(1), 82–86 (2005)
46. O.S.V. Ehrenstein, D.N.G. Mazumder, M.H. Smith, N. Ghosh, Y. Yuan, G. Windham, A. Ghosh, R. Haque, S. Lahiri, D. Kalman, S. Das, A.H. Smith, Pregnancy outcomes, infant mortality, and arsenic in drinking water in West Bengal, India. Am. J. Epidemiol. **163**, 662–669 (2006)
47. R.K. Kwok, R.B. Kaufmann, M. Jakariya, Arsenic in drinking-water and reproductive health outcomes: a study of participants in the Bangladesh integrated nutrition programme. J. Health Popul. Nutr. **24**(2), 190–205 (2006)

48. C. Hopenhayn, C. Ferreccio, S.R. Browning, B. Huang, C. Peralta, H. Gibb, I. Hertz-Picciotto, Arsenic exposure from drinking water and birth weight. Epidemiology **14**, 593–602 (2003)
49. C.Y. Yang, C.C. Chang, S.S. Tsai, H.Y. Chuang, C.K. Ho, T.N. Wu, Arsenic in drinking water and adverse pregnancy outcome in an arseniasis-endemic area in north eastern Taiwan. Environ. Res. **91**(1), 29–34 (2003)
50. M.L. Kile, A. Cardenas, E. Rodrigues, M. Mazumdar, C. Dobson, M. Golam, Q. Quamruzzaman, M. Rahman, D.C. Christiani, Estimating effects of arsenic exposure during pregnancy on perinatal outcomes in a Bangladeshi cohort. Epidemiology **27**, 173–181 (2016)
51. K.L. Huyck, M.L. Kile, G. Mahiuddin, Q. Quamruzzaman, M. Rahman, C.V. Breton, C.B. Dobson, J. Frelich, E. Hoffman, J. Yousuf, S. Afroz, S. Islam, D.C. Christiani, Maternal arsenic exposure associated with low birth weight in Bangladesh. J. Occup. Environ. Med. **49**, 1097–1104 (2007)
52. C. Hopenhayn, K.D. Johnson, I. Hertz-Picciotto, Reproductive and developmental effects associated with chronic arsenic exposure. Presented at the 3rd international conference on arsenic exposure and health effects, San Diego, CA, 12–15 July 1998
53. S. Nordstrom, L. Beckman, I. Nordenson, Occupational and environmental risk in and around a smelter in northern Sweden—III. Frequencies of spontaneous abortion. Hereditas **88**, 51–54 (1978)
54. S. Nordstrom, L. Beckman, I. Nordenson, Occupational and environmental risk in and around a smelter in northern Sweden—V. Spontaneous abortion among female employees and decrease birth weight in their offspring. Hereditas **90**, 291–296 (1979)
55. S. Nordstrom, L. Beckman, I. Nordenson, Occupational and environmental risks in and around a smelter in northern Sweden—VI. Congenital malformations. Hereditas **90**, 297–302 (1979)
56. S. Tabacova, D.D. Baird, L. Balabaeva, D. LoLova, I. Petrov, Placental arsenic and cadmium in relation to lipid peroxides and glutathione levels in maternal-infant pairs from a copper smelter area. Placenta **8**, 873–881 (1994)
57. C.T. Bolliger, P. Van Zigl, J.A. Louw, Multiple organ failure with adult respiratory di stress syndrome in homicidal arsenic poisoning. Respiration **59**, 57–61 (1992)
58. G. Lugo, G. Cassady, P. Palmisano, Acute maternal arsenic intoxication with neonatal death. Am. J. Dis. Child. **117**, 328–330 (1969)
59. A. Rahman, M. Vahter, E.C. Ekström, M. Rahman, A.H. Golam Mustafa, M.A. Wahed, M. Yunus, L.A. Persson, Association of arsenic exposure during pregnancy with fetal loss and infant death: a cohort study in Bangladesh. Am. J. Epidemiol. **165**(12), 1389–1396 (2007)
60. S.L. Myers, D.T. Lobdell, Z. Liu, Y. Xia, H. Ren, Y. Li, R.K. Kwok, J.L. Mumford, P. Mendola, Maternal drinking water arsenic exposure and perinatal outcomes in Inner Mongolia. China J. Epidemiol. Community Health **64**, 325–329 (2009)
61. L. Beckman, S. Nordstrom, Occupational and environmental risks in and around a smelter in Northern Sweden. IX. Fetal mortality among wives of smelter workers. Heriditas **97**, 1–7 (1982)
62. S. Zierler, M. Theodore, A. Cohen, K.J. Rothman, Chemical quality of maternal drinking water and congenital heart disease. Int. J. Epidemiol. **17**, 589–594 (1988)
63. A. Aaschengrau, S. Zierler, A. Cohen, Quality of community drinking water and the occurrence of spontaneous abortion. Arch. Environ. Health **44**, 283–290 (1989)
64. M. Börzsönyi, A. Bereczky, P. Rudnai, M. Csanady, A. Horvath, Epidemiological studies on human subjects exposed to arsenic in drinking water in southeast Hungary. Arch. Toxicol. **66**(1), 77–78 (1992)
65. M.S. Bloom, S. Surdu, L.A. Neamtiu, E.S. Gurzau, Maternal arsenic exposure and birth outcomes: a comprehensive review of the epidemiologic literature focused on drinking water. Int. J. Hyg. Environ. Health **217**, 709–719 (2014)
66. M.S. Bloom, E.F. Fitzgerald, K. Kim, L. Neamtiu, E.S. Gurzau, Spontaneous pregnancy loss in humans and exposure to arsenic in drinking water. Int. J. Hyg. Environ. Health **213**(6), 401–413 (2010)
67. R. Raqib, S. Ahmed, R. Sultan, Y. Wagatsum, D. Mondal, A.M. Waheedul Hoque, B. Nermell, M. Yunus, S. Roya, L. Ake Perssone, S.E. Arifeena, S. Moore, M. Vahter, Effects of in utero arsenic exposure on child immunity and morbidity in rural Bangladesh. Toxicol. Lett. **185**, 197–202 (2009)

68. S. Ahmed, S. Mahabbat-e Khoda, R.S. Rekha, R.M. Gardner, S.S. Ameer, S. Moore, E.-C. Ekström, M. Vahter, R. Raqib, Arsenic-associated oxidative stress, inflammation, and immune disruption in human placenta and cord blood. Environ. Health Perspect. **119**, 258–264 (2011)
69. M. Vahter, Effects of arsenic on maternal and fetal health. Annu. Rev. Nutr. **29**, 381–399 (2009)
70. C.H. Tseng, A review on environmental factors regulating arsenic methylation in humans. Toxicol. Appl. Pharmacol. **235**, 338–350 (2009)
71. W.C. Chou, Y.T. Chung, H.Y. Chen, C.J. Wang, T.H. Ying, C.Y. Chuang, Y.H. Tseng, S.L. Wang, Maternal arsenic exposure and DNA damage biomarkers, and the associations with birth outcomes in a general population from Taiwan. PLoS One **9**(2), e86398 (2014)
72. J.R. Pilsner, M.N. Hall, X. Liu, V. Ilievski, V. Slavkovich, D. Levy, P.M. Litvak, M. Yunus, M. Rahman, J.H. Graziano, M.V. Gamble, Influence of prenatal arsenic exposure and newborn sex on global methylation of cord blood DNA. PLoS One **7**(5), e37147 (2012)
73. H. Chen, J. Liu, B.A. Merrick, M.P. Waalkes, Genetic events associated with arsenic-induced malignant transformation: applications of cDNA microarray technology. Mol. Carcinog. **30**(2), 79–87 (2001)
74. J.F. Reichard, A. Puga, Effects of arsenic exposure on DNA methylation and epigenetic gene regulation. Epigenomics **2**(1), 87–104 (2010)
75. P. Intarasunanont, P. Navasumrit, S. Woraprasit, K. Chaisatra, W.A. Suk, C. Mahidol, M. Ruchirawat, Effects of arsenic exposure on DNA methylation in cord blood samples from newborn babies and in a human lymphoblast cell line. Environ. Health **11**(1), 31 (2012)
76. D.C. Koestler, M. Avissar-Whiting, E.A. Houseman, M.R. Karagas, C.J. Marsit, Differential DNA methylation in umbilical cord blood of infants exposed to low levels of arsenic in utero. Environ. Health Perspect. **121**, 971–977 (2013)
77. M.T. Islam, S. Parvin, M. Pervin, A.S.M. Bari, M. Khan, Effects of chronic arsenic toxicity on the haematology and histoarchitecture of female reproductive system of black Bengal goat. Bangl. J. Vet. Med. **9**(1), 59–66 (2011)
78. S. Chattopadhyay, S. Ghosh, S. Chaki, J. Debnath, D. Ghosh, Effect of sodium arsenite on plasma levels of gonadotrophins and ovarian steroidogenesis in mature albino rats: duration dependent response. J. Toxicol. Sci. **24**, 425–431 (1999)
79. P.A. Navarro, L. Liu, D.L. Keefe, In vivo effects of arsenite on meiosis, preimplantation development, and apoptosis in the mouse. Biol. Reprod. **70**, 980–985 (2004)
80. C. Zhang, B. Ling, J. Liu, G. Wang, Toxic effect of fluoride-arsenic on the reproduction and development of rats. J. Hyg. Res. **29**, 138–140 (2000)
81. M.S. Golub, M.S. Macintosh, N. Baumrind, Developmental and reproductive toxicity of inorganic arsenic: animal studies and human concerns. J. Toxicol. Environ. Health B Crit. Rev. **1**(3), 199–241 (1998)
82. H.D. Hood, Effects of sodium arsenite on fetal development. Bull. Environ. Contamin. Toxicol. **7**(4), 216–222 (1972)
83. F. Holson, G. Stump, J. Clevidence, J.F. Knapp, H. Farr, Evaluation of the prenatal developmental toxicity of orally administered arsenic trioxide in rats. Food Chem. Toxicol. **38**, 459–466 (2000)
84. Z. Akram, S. Jalali, S.A. Shami, L. Ahmad, S. Batool, O. Kalsoom, Adverse effects of arsenic exposure on uterine function and structure in female rat. Exp. Toxicol. Pathol. **62**(4), 451–459 (2010)
85. A. Chatterjee, U. Chatterji, Arsenic abrogates the estrogen-signaling pathway in the rat uterus. Reprod. Biol. Endocrinol. **8**, 80 (2010)
86. S. Chattopadhyay, D. Ghosh, Role of dietary GSH in the amelioration of sodium arsenite-induced ovarian and uterine disorders. Reprod. Toxicol. **30**, 481–488 (2010)
87. S. Chattopadhyay, S. Pal, D. Ghosh, J. Debnath, Effect of dietary co administration of sodium selenite on sodium arsenite-induced ovarian and uterine disorders in mature albino rats. Toxicol. Sci. **75**, 412–422 (2003)
88. J.N. Treas, T. Tyagi, K.P. Singh, Effects of chronic exposure to arsenic and estrogen on epigenetic regulatory genes expression and epigenetic code in human prostate epithelial cells. PLoS One **7**(8), e43880 (2012)

89. D. Montalbano, D. Jacobson-Kram, The reproductive effects assessment group's report on the mutagenicity of inorganic arsenic. Environ. Mutagen. **7**, 787–804 (1985)

90. C.D. Kozul-Horvath, F. Zandbergen, B.P. Jackson, R.I. Enelow, J.W. Hamilton, Effects of low-dose drinking water arsenic on mouse fetal and postnatal growth and development. PLoS One **7**(5), e38249 (2012)

91. M.P. Waalkes, L.K. Keefer, B.A. Diwan, Induction of proliferative lesions of the uterus, testes, and liver in Swiss mice given repeated injections of sodium arsenate: possible estrogenic mode of action. Toxicol. Appl. Pharmacol. **166**, 24–35 (2000)

92. W. He, R.J. Greenwell, D.M. Brooks, L. Calderon-Garciduenas, H.D. Beall, J.D. Coffin, Arsenic exposure in pregnant mice disrupts placental vasculogenesis and causes spontaneous abortion. Toxicol. Sci. **99**, 244–253 (2007)

93. B. Patel, R. Das, A. Gautam, M. Tiwari, S. Acharya, S. Kumar, Evaluation of vascular effect of arsenic using in vivo assays. Environ. Sci. Pollut. Res. (Under publication) (2017)

94. A.S. Ettinger, A.R. Zota, C.J. Amarasiriwardena, M.R. Hopkins, J. Schwartz, H. Hu, R.O. Wright, Maternal arsenic exposure and impaired glucose tolerance during pregnancy. Environ. Health Perspect. **117**(7), 1059–1064 (2009)

95. M.P. Reilly, J.C. Saca, A. Hamilton, R.F. Solano, J.R. Rivera, W. Whitehouse-Innis, J.C. Parsons, R.L. Dearth, Prepubertal exposure to arsenic(III) suppresses circulating insulin-like growth factor-1 (IGF-1) delaying sexual maturation in female rats. Reprod. Toxicol. **44**, 41–49 (2014)

96. P.H. Ser, B. Banu, F. Jebunnesa, K. Fatema, N. Rosy, R. Yasmin, H. Furusawa, L. Ali, S.A. Ahmad, C. Watanabe, Arsenic exposure increases maternal but not cord serum IgG in Bangladesh. J. Pediatr. Int. **57**(1), 119–125 (2015)

97. M.A. Wares, M.A. Awal, S.K. Das, J. Alam, Environmentally persistant toxicant arsenic affects uterus grossly and histologically. Bangl. J. Vet. Med. **11**(1), 61–68 (2013)

98. A. Hedayati, A.A. Sadeghi, P. Shawrang, The ameliorative effect of catechin on sexual hormones in female rats exposed to arsenic toxicity. Indian J. Fundam. Appl. Life Sci. **5**(1), 238–243 (2015)

99. S. Mondal, S. Mukherjee, K. Chaudhuri, S.N. Kabir, P.K. Mukhopadhyay, Prevention of arsenic-mediated reproductive toxicity in adult female rats by high protein diet. Pharm. Biol. **51**(11), 1363–1371 (2013)

Chapter 6
Fish, Genetic and Cellular Toxicity

Fluorescence in situ hybridization (FISH) is a cytogenetic technique used to detect and localize the presence or absence of specific DNA sequences on chromosomes. FISH uses fluorescent probes bind to those targets that show a high degree of sequence complementarity, and it can help define the spatial-temporal patterns of gene expression within cells and tissues. Comet assay and micronucleus (MN) test are used in genotoxicity studies and testing. This technique permits to measure direct DNA-strand breaking capacity of a tested agent, and MN test allows estimating the details of genome mutations. FISH and comet assay help in identification of DNA damage and repairing. This paper lists relevant material on advantages and limitations of Comet-FISH and MN-FISH assays application in genetic toxicology and cellular toxicity studies. This is useful in study of whole DNA damage repair and genotoxicity exposure and effect studies too [1].

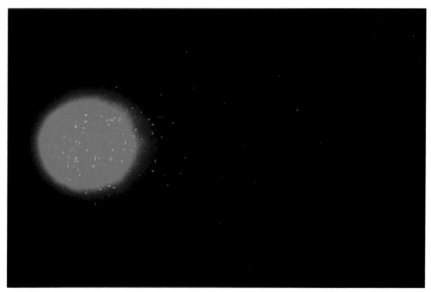

Example of SYBR-green-stained comet image from BLM-treated human leuko-cytes with telomeric PNA probes indicating the location of telomeric repeat sequences [1].

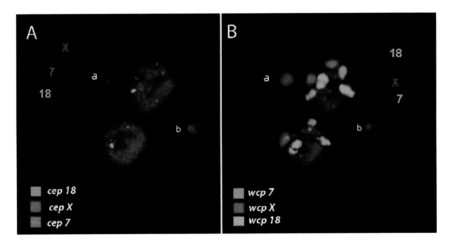

Example of DAPI-stained binucleated cell image from MMC-treated human leukocytes with centromeric (A) and whole chromosome painting (B) probes for chromosomes 7, 18 and X [1].

Cellular toxicity or cytotoxicity is the phenomenon of cell destruction or lysis in response to either the metabolic changes or the external agents. In this case, the cell gets damaged or lysed following either of the two major phenomena cell necrosis and apoptosis.

Cell necrosis refers to the killing of the cells accidentally by microbes or chemical agents. This is unplanned and sudden cell loss. In case of necrosis, the cells are killed by other immune cells or medicated by a chemical agent. Necrosis is a responsive action to a particular stimulus.

Apoptosis refers to the programmed cell death which leads to the renewal of the fresh and new cells after every cycle. The older and damaged cells are subjected to self-destruction to give chance for the new cells to rejuvenate.

6.1 Cytotoxic Agents

According to the NCI dictionary a cytotoxic agent is a substance that kills the cells including the cancer cells by causing the cells to shrink and die [2].

Two major agents causing cytotoxicity are immune cells and chemicals

1. **Immune cells in inducing cytotoxicity**

Each organism is gifted with specialized defense cells that can cause the cell cycle disruption or cell lysis of the target cells which are either toxic to the body or foreign in nature. These immune cells cause lysis of such cells as a means to protect the body and provide immunity. Some of the cytotoxic immune cells include T cells, natural killer cells, lymphokine-activated killer cells, macrophages, lysosomal granules, MHC complex, etc.

All the different cytotoxic cells differ in their morphology, origin, phenotype and target cell specificity. However the mechanism by which these cells causes lysis of foreign cells is by contact-dependent non-phagocytic process.

2. **Chemicals and drugs involved in cytotoxicity**

All the chemicals, derivatives and drugs that can induce cell destruction or lysis can be termed as cytotoxic chemicals. Most of the drugs used in the treatment of cancer come under cytotoxic chemicals/drugs. They cause cell lysis by blocking the cell cycle or stopping the cell division. These chemicals may also be cytostatic, causing an obstruction for cell division. Most of the known cytotoxic agents are known to possess the properties of carcinogenic, teratogenic, toxic and reproductive effects, acute effects and chronic effects [3].

Carcinogenic Use of cytotoxic agents in therapeutic doses may in times lead to the formation of malignant tumors. Cyclosporine and cyclophosphamide fall in the list and are declared to be carcinogens by NTP and IARC.

Teratogenic All the chemicals and agents which cause a developmental abnormalities leading to physical impairment of the effected fetus come in this category. Cyclophosphamide is a medication used to suppress the immune system. It often exhibits teratogenic effect.

Toxic and reproductive effects Some of these effects include reduced sperm count, lack of sperm production, damage to the bone marrow, amenorrhea, affects the heart, kidneys, nervous system, etc. Anti-epileptic medicine like carbamazepine is shown to be detrimental for sperm morphology and functioning.

Acute effects Some of the cytotoxic chemicals are very strong and can induce cellular damage upon mere contact. The may act as vesicants causing local tissue necrosis. Dactinomycin, daunorubicin, doxorubicin are all cytotoxic agents having the vesicant properties and causing tissue necrosis.

Chronic effects A study conducted by Selevan et al. reveled that those nurses who are constantly exposed to cytotoxic agents later developed genetic problem leading to fetal death. Several other chronic effects include liver damage, skin damage, nausea, giddiness, etc. For example, busulfan and bleomycin may cause lung infection upon long term use.

6.2 Factors and Mechanism of Cell Toxicity

Cell toxicity is a major phenomenon that may lead to cellular stress, damage and death. Some of the major factors leading to the cell toxicity are:

1. Excess nitric oxide production (NO)
2. Oxidative stress by reactive oxygen species (ROS)
3. Mitochondrial dysfunction due to oxidative stress
4. DNA damage.

1. **High concentration of NO production in the cell**

The presence of NO in the cell is both beneficial and detrimental. This is an essential element in the proper functioning of endothelial cells. The regular vascular flow is maintained by the proper concentration of NO in the cells. However the elevated levels of NO in the cell causes oxidative stress and cellular burst [4]. This high concentration of NO is toxic to the cell and may lead to cell damage. One of the examples is doxorubicin which induces disfunctioning of skeletal muscles due to the intermediate release of NO. This is also known to disrupt the functions of phagocytes. Several cellular pathways may involve in the overproduction of NO leading to cell toxicity.

In some conditions like immune responses by the cells, the body responds by releasing excess NO. Accumulation of this NO in times causes cell death or cytotoxicity.

Additionally, excess concentration of NO in the cell may lead to the following.

Suppression of NF-κB pathway induces neurotoxicity mediated by TNF alpha cells and activates p38 MAPK and p53 pathway. It also causes the fragmentation of DNA helix in the microglia. Thus, increased NO in the cell may lead to several abnormalities finally causing cytotoxicity.

2. **Oxidative stress by reactive oxygen species**

Reactive oxygen species are the elements with highly reactive oxygen with them and possess very low stability. The presence of active oxygen imparts a high degree of oxidative stress. Metabolic processes like increased metabolic activity, mitochondrial dysfunction, peroxisome activity, increased cellular receptor signaling, etc., are the major reasons behind the elevated levels of ROS. These elevated levels of ROS further result in an extensive damage to the DNA, proteins and lipids, finally causing cytotoxicity. Some of the ROS species are superoxide anion (O_2^-), hydrogen peroxide (H_2O_2), and hydroxyl radicals ($OH^.$).

ROS is a very sensitive aspect in imparting immunity to the organisms. Maintenance of homeostasis in ROS is essential for the regular functioning of the cell. Various chemicals and drugs administered may lead to the increased production in the cellular ROS leading to cytotoxic effects. One of the examples for ROS induced cell toxicity is the use of acetaminophen on liver cells. The treated liver cells exhibited increased concentrations of ROS with a simultaneous depletion of glutathione resulting in liver cell death. Another example is the chemical pentachlorophenol

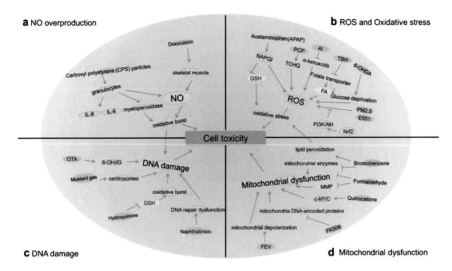

a NO overproduction

b ROS and Oxidative stress

c DNA damage

d Mitochondrial dysfunction

Fig. 6.1 Factors and mechanism of cytotoxicity. *Pic courtesy* PMID: 30374795

which is a pesticide upon I reaction with the body causes an increased production of ROS and death of hepatocytes.

3. **Mitochondrial dysfunction due to oxidative stress**

Mitochondria being the power houses of the cell are very sensitive and undergo damage due to the oxidative stress caused by ROS leading to cellular dysfunction and contributing to the cytotoxicity.

4. **DNA damage due to ROS**

The increased concentration of reactive oxygen species especially the mitochondrial ROS causes an intense pressure on the mitochondria leading to a variety of DNA damages like oxidization of bases and breakage in the double helical strands. This causes an ultimate damage to mitochondria. Under normal cellular conditions, the ROS generated are always regulated in concentration by maintaining the scavenging system. However under the oxidative stress, the cells ability to maintain the balance is lost causing damage to the cells. The damage of the genomic DNA due to the oxidative stress generated by ROS leads to the damage of the cell (Fig. 6.1).

6.3 Biomarkers in Cell Toxicity Study

In view of the cellular toxicity caused by several drugs and chemicals used in therapy, it becomes an important aspect to have preliminary information regarding the degree of risk or toxicity involved while using a selected chemical/drug [5]. Thus, risk assessment in the use of drugs is a subject of importance. Use of biomarkers for the assessment of risk is proved to be beneficial.

Biomarker may be a natural substance or a metabolite present in the cell whose tracing can detail the effect of a chemical in the body. These markers yield information about the state of health of an individual.

Some of the natural phenomenon occurring in the cell may be selected as biomarkers as described below:

1. **Autophagy: define autophagy with diagram**

Autophagy is the natural phenomenon of the body to eliminate old and damaged cells. Here, the cells are engulfed by the immune cells [6]. Thus, the term autophagy which means self-killing is used. It is further classified into micro-autophagy, macro-autophagy and chaperon-mediated autophagy based on the type of immune cells involved in the process and the size of the cells being destroyed.

Autophagy is also a natural response of body toward cancer therapy; thus, the degree of autophagy can be used as an efficient biomarker for the prediction of anticancer efficiency of drugs.

2. **ROS as a marker for predicting cell toxicity**

Increase in the production of cellular ROS is an indication of cell toxicity. The presence of toxic agents causes an increase in the oxidative stress leading to an overproduction of ROS. This in turn causes cell toxicity [7]. An increase in the ROS acts as direct marker for the increased cell toxicity.

3. **Cytokines in cell toxicity**

Cytokines are small protein molecules present on the cells and involved in cell signaling. Being cell signaling molecules they lack the capability to enter into the cell and are only localized to the cell surface [8]. They play a major role in autocrine, paracrine and endocrine signaling. They act as immune modulators. In some cases of influenza-associated diseases, elevated levels of cytokines can be used as markers. Several groups of cytokines may act as markers for inflammation induced acute respiratory disease syndrome (ARDS). These biomarkers include bone morphogenetic protein-15, CXCL16, CXCR3, IL-6, protein NOV homolog, glypican 3, IGFBP-4, IL-5, IL-5R alpha, IL-22 BP, leptin, MIP-1d and orexin B. The intensity and degree of disease is indicated by the elevated levels of IL-6, CXCL16 or IGFBP-4.

6.4 Measuring the Cell Toxicity

There are several ways to measure the cell toxicity, one of them being measurement of cell viability [9]. This can be done using certain dyes like formalin. Protease biomarkers, ATP content measurement are some of the methods. Some of the chromogenic dyes like formazan would indicate the cell toxicity by the development of a typical color. SRB and WST-1 assays are considered to be ideal for the measurement of cytotoxicity.

Viability count of the cells is the number of cells in live and healthy condition. This testing is performed using some special dyes. The cell viability testing is based on several factors like: enzyme activity, cell membrane permeability, cell adherence, ATP production, co-enzyme production and nucleotide uptake activity. There are several protocols for the cell viability measurement like

1. **DNA Synthesis and Cell Proliferation Assay**: This protocol uses 3H thymidine having radioactivity to interact with the cells overnight [10]. All the actively proliferating cells incorporate the radioactive dye in their DNA. This can later be washed and tested for the active cells using scintillation counter (Fig. 6.2).
2. **Metabolic Cell Proliferation Assay**: The metabolic activity of the cells is one of the direct analyses of the population of cells [11]. Tetrazolium salts or Alamar Blue are the compounds that indicate the metabolic state of the cell by getting reduced. Upon reduction, the native color of the media changes due to the formation of formazan dye indicating the proliferation of the cell. The complete reaction depends on the production of lactate dehydrogenase enzyme produced during cell division. The color developed is read using a spectrophotometer or micro-plate reader.

   MTT, XTT, MTS and WST1 are the commonly used salts for analysis of the cell proliferation [12] (Fig. 6.3).
3. **Measuring the ATP Content in the Cell**: As the concentration of ATP is an indication of cell viability and function, ATP content of the cell is read using luciferase: luciferin bioluminescent enzyme couple. Here in the reaction enzyme

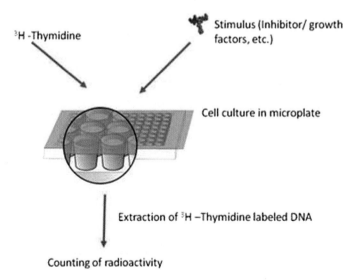

Fig. 6.2 Measurement of ATP content in the cell. *Pic courtesy* PerkinElmer

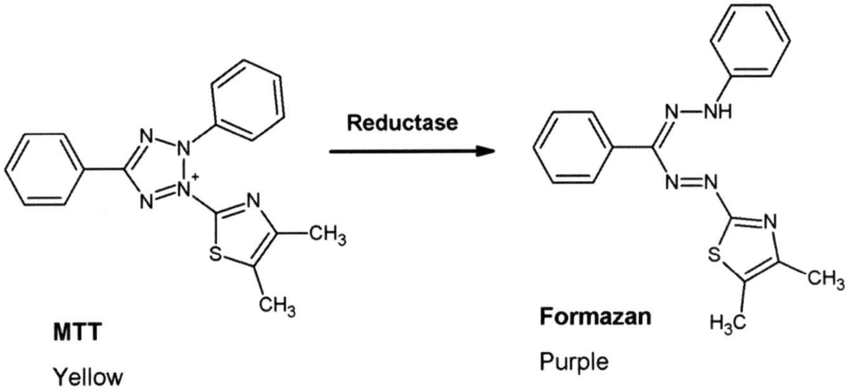

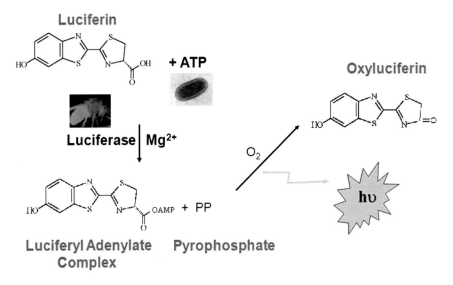

Fig. 6.3 *Pic courtesy* Wolfgang E. Trommer

Luciferase produces light upon reacting with luciferin in the presence of ATP [13]. The intensity of the light released is proportional to the amount of ATP generated which is read using a luminometer or any other micro plate reader which can detect the luminescent signals (Fig. 6.4).

Fig. 6.4 Reaction. *Pic courtesy* Creative bioarray

References

1. G.G. Hovhannisyan, Fluorescence in situ hybridization in combination with the comet assay and micronucleus test in genetic toxicology. Mol. Cytogenet. **3**, 17 (2010). https://doi.org/10. 1186/1755-8166-3-17
2. *NCI Dictionary of Cancer Terms*, National Cancer Institute, NIH
3. EXTRACNET, Environmental Health and Safety, Laboratory and chemical hygiene, Chap. 3, in *12.0 Chemical Hazard Classes and Special Instructions*
4. Y. Zhang, Cell toxicity mechanism and biomarker. Clin. Transl. Med. (2018). PMID: 30374795
5. S.G. Mina et al., Assessment of drug-induced toxicity biomarkers in the brain microphysiological system (MPS) using targeted and untargeted molecular profiling. Front. Big Data Med. Public Health (2019)
6. A. Thorburn et al., Autophagy and cancer therapy. Mol. Pharmacol. **85**(6), 830–838 (2014)
7. H. Furue, Toxicity criteria. Gan To Kagaku Ryoho **22**(5), 616–626 (1995)
8. S. Chiang, D.S. Kalinowski, P.J. Jansson, D.R. Richardson, M.L. Huang, Mitochondrial dysfunction in the neuro-degenerative and cardio-degenerative disease, Friedreich's ataxia. Neurochem. Int. **117**, 35–48 (2017). https://doi.org/10.1016/j.neuint.2017.08.002
9. T. Decker, M.L. Lohmann-Matthes, A quick and simple method for the quantitation of lactate dehydrogenase release in measurements of cellular cytotoxicity and tumor necrosis factor (TNF) activity. J. Immunol. Methods **115**(1), 61–69 (1988). Epub 1988/11/25
10. A. Adan, Y. Kiraz, Y. Baran, Cell proliferation and cytotoxicity assays. Curr. Pharm. Biotechnol. **17**(14), 1213–1221 (2016)
11. L.J. Jones et al., Sensitive determination of cell number using the CyQUANT® cell proliferation assay. J. Immunol. Methods **254**(1–2) (2001)
12. M. Ginouves et al., Comparison of tetrazolium salt assays for evaluation of drug activity against *Leishmania* spp. J. Clin. Microbiol. **52**(6), 2131–2138 (2014)
13. Ö.S. Aslantürk, In vitro cytotoxicity and cell viability assays: principles, advantages, and disadvantages, in *Genotoxicity—A Predictable Risk to Our Actual World*, ed. by M.L. Larramendy, S. Soloneski (IntechOpen, 2017). https://doi.org/10.5772/intechopen.71923

Printed in the United States
By Bookmasters